# 1st International Congress on Sustainability Science and Engineering (ICOSSE 2009)



Cincinnati, Ohio, USA
9 - 13 August 2009

ISBN: 978-1-5108-8548-6

**Printed from e-media with permission by:**

Curran Associates, Inc.
57 Morehouse Lane
Red Hook, NY 12571

Some format issues inherent in the e-media version may also appear in this print version.



Printed by Curran Associates, Inc. (2019)

For permission requests, please contact AIChE
at the address below.

AIChE
120 Wall Street, FL 23
New York, NY 10005-4020

Phone: (800) 242-4363
Fax: (203) 775-5177

www.aiche.org

**Additional copies of this publication are available from:**

Curran Associates, Inc.
57 Morehouse Lane
Red Hook, NY 12571 USA
Phone: 845-758-0400
Fax: 845-758-2633
Email: curran@proceedings.com
Web: www.proceedings.com

# TABLE OF CONTENTS

## INVITED PLENARY SESSION: PERSPECTIVE OF INDUSTRIAL SUSTAINABILITY OUTCOMES



## SUSTAINABILITY AS A SCIENTIFIC PARADIGM FOR SOLUTIONS



## MEASURING SUSTAINABILITY



## STUDENT POSTER SESSION

## VALUE/SUPPLY CHAIN SUSTAINABILITY



## SOLUTIONS FOR FOSSIL AND NON-FOSSIL POWER

## SOLUTIONS FOR SUSTAINABLE TRANSPORTATION



## SUSTAINABLE MATERIAL/PRODUCT DESIGN/MANUFACTURING



## SPECIAL DISPLAY SESSION

## SUSTAINABLE PROCESSES/LCA ENGINEERING



## MODELLING—DESIGNING FOR SUSTAINABILITY AND DECISION MAKING



## CODES/STANDARDS AND SCIENCE-BASED GUIDANCE FOR SUSTAINABILITY

## SUSTAINABLE DESIGN AND ARCHITECTURE

# The Sustainability of International Supply Chains: Aspirations and Practicality

## Authors:

Clift, R., University of Surrey

Sustainability has come to be interpreted widely in terms of three ?pillars? or constraints: techno-economic, environmental or ecological, and societal. Sustainable development then means improving quality of life for all, within these constraints ? in the words of Professor Tim Jackson, ?Sustainability is the art of living well within ecological limits? where ?living well? implies ethical behaviour as well as comfort.
These principles should be applied to international supply chains from a perspective which sees social and economic benefit along with pollution and resource use as resulting primarily from consumption rather than production. This leads to the idea that the sustainability of supply chains should be assessed by examining the distribution of positive benefits and negative impacts along the supply chain. The economic dimension can be addressed by Value Chain Analysis and the environmental dimension by Life Cycle Assessment; juxtaposing the results from VCA and LCA immediately gives useful insights into the sustainability of supply chains. Tools for assessing the distribution of social benefits along a supply chain are still in development but are equally important in a market where consumers and, even more significantly, retailers require suppliers to demonstrate Corporate Social Responsibility.

This approach to assessing sustainability has been explored by examining some specific supply chains. The results show both the possibilities and the challenges of attempting to improve the sustainability of international supply chains.

**Chemical & Allied Industry Perspective**

Madzy, E. G., BASF Corp.

Originally presented on: 8/10/2009 9:10:00 - 9:50:00

Sustainability is a 21st century business imperative. The chemicalindustry can supply solutions to the pressing issues our global society faces today and will face in the future. At BASF, "Ensure Sustainable Development" is one of four core principles that guide all business strategies and decisions. BASF's leadership on sustainability is supported by an array of sophisticated tools that help identify societal needs and generate innovative market solutions aimed at improving the quality of life today without jeopardizing the opportunities of future generations. Advances in the area of energy, construction, mobility,health and nutrition have propelled BASF to a leadership role in the chemical industry. It is incumbent upon the chemical industry to clearly demonstrate that, where climate protection and energy management is concerned, we are a part of the solution not a part of the problem. Notions to the contrary are based on a paradigm that is three decades outdated and must be dispelled through continual, demonstrable industry-wide commitment to sustainable practices and products.

**Sustainability at Procter & Gamble**
Sauers, L., Procter & Gamble
Originally presented on: 8/10/2009 9:50:00 - 10:30:00

At P&G, sustainability is defined broadly. It is about ensuring a better quality of life for everyone now and for generations to come. It is a broad definition that encompasses both environmental and social responsibility. For us, it's not just about ?green? products tied to a current consumer movement, but a systemic approach covering our total supply chain, from raw materials to manufacturing processes to the consumer.

Our disciplined, scientific approach to sustainability includes Life Cycle Assessment (LCA). P&G helped pioneer this research tool which looks at environmental factors such as carbon dioxide, energy, water consumption, and waste over the entire lifespan of a product from the creation of raw materials, to product manufacture and logistics, to consumer use, and to the final post-consumer disposal. This comprehensive approach helps identify the biggest opportunities to improve the environmental impact of products.

P&G's sustainability program is focused in 5 key areas:

(1) Improving the environmental profile of our products

(2) Improving the environmental profile of our operations

(3) Advancing social responsibility via our Live, Learn and Thrive program

(4) Engaging our employees

(5) Partnering with external stakeholders

This session will review P&G's sustainability program, with emphasis on how it has evolved from being one focused only on corporate responsibility to additionally seeing sustainability as an opportunity for business growth.

**Sustainable It Ecosystem: Enabling City Scale Infrastructures**
Patel, C., Hewlett Packard
Originally presented on: 8/10/2009 10:50:00 - 11:30:00

Continued population growth coupled with increased per capita consumption will soon require the creation of a new generation of cities in emerging economies and around the globe. But unlike the previous generation, "City 2.0" will be driven by a growing desire for on-demand, just-in-time access to critical resources at affordable costs. Coupled with the increased risks that physical growth places on the carrying capacity of the biosphere, we cannot expect to meet the future needs of society simply by extending the existing infrastructures.

Instead, in this talk, I propose that the necessary transformation be enabled by a global IT ecosystem made up of billions of service-oriented client devices and thousands of data centers. The IT ecosystem, with data centers at its core, and pervasive measurement at the edges, will need to be seamlessly integrated into future communities to enable need-based provisioning of critical resources. Such a transformation of the IT ecosystem necessitates addressing sustainability from a cradle-to-cradle perspective ? minimizing the available energy needed for extraction, manufacturing, waste mitigation, transportation, operation and reclamation. This lifecycle based engineering and management will result in achieving a reduced Total Cost of Ownership (TCO) that will serve the billions.

**Automotive Industry Perspective**

Taub, A., General Motors

Originally presented on: 8/10/2009 11:30:00 - 12:10:00

The auto industry today is reinventing the automobile using a new ?DNA? that will make our future vehicles more sustainable in terms of energy, the environment, safety, congestion, and affordability. This new automotive DNA will encompass a range of technology solutions. In addition to electrification of the propulsion system and vehicles powered from renewable energy sources, we will also see new telematics and ?connected vehicle? capabilities, advanced on-board electronics, lightweight and novel materials, and innovative vehicle designs and aerodynamics. In this talk, Dr. Taub will detail the most promising technology options and discuss the major opportunities and issues in each area. He will also highlight why the new DNA will revolutionize our vehicles, be paradigm shifting for the industry, and help move the world to a more sustainable future.

# Visions for Sustainable Infrastructure

## Authors:

Boyle, C., University of Auckland

Unprecedented populations in cities have forced some infrastructure systems to become larger, more complex and more interdependent. The decisions that are made today on delivering services to cities can have long term consequences, affecting the resilience, economics and resources of the city in the future as well as the health and wellbeing of city inhabitants. New technologies are also changing the way we can deliver infrastructure services, thus redefining some of our basic premises about infrastructure needs and creating clashes between old, now inefficient but built in technologies and new, completely different and much more efficient technologies. In addition, infrastructure, including the built environment, is a major consumer of funding, energy and material resources, during both construction and maintenance. Yet healthy and functioning infrastructures are the key to a sustainable city as they deliver the essential services to fulfil the needs of both current and future generations. A new vision for sustainable infrastructure is needed, which will recognise the long term future of our cities, the need to deliver basic services to all inhabitants and the implications for businesses and manufacturers. This paper presents some of the elements which will have to be taken into consideration in that new vision.

**Fuel and Vehicle Technology Choices for Passenger Vehicles in a Sustainable World**

Wallington, T. J., Ford Motor Company
Anderson, J. E., Ford Motor Company
Mueller, S. A., Ford Motor Company
Grahn, M., Chalmers University of Technology
Williander, M. I., Chalmers University of Technology
Originally presented on: 8/10/2009 14:10:00 - 14:35:00

The multi-sector regionalized Global Energy Transition (GET-RC 6.1) model covers the global energy system and is designed to meet exogenously-given energy demand levels while stabilizing at a specific atmospheric CO2 concentration at the lowest system cost. Energy demand is aggregated into three sectors: (i) electricity, (ii) transportation, and (iii) "heat" which comprises all stationary uses of energy (industry, commercial, residential) except those associated with providing electricity or transportation fuels. The model includes passenger and freight transportation by air, road, rail, and water. We have updated the model to include a detailed description of fuel and vehicle technologies for passenger vehicles. Five fuel options (petroleum, natural gas, synthetic fuels [coal to liquid, CTL; gas to liquid, GTL; biomass to liquid, BTL], electricity, and hydrogen) were considered. Five vehicle powertrain technologies (internal combustion engine vehicles [ICEVs], hybrid electric vehicles [HEVs], plug-in hybrid electric vehicles [PHEVs], battery electric vehicles [BEVs], and fuel cell vehicles [FCVs]) were included. Carbon capture and storage (CCS) and concentrating solar power (CSP) were available in the model as examples of low-CO2 electricity and/or hydrogen technologies. Passenger vehicle technology and fuel choices consistent with stabilization of atmospheric CO2 at 400-550 ppm by 2100 at the lowest total system cost were investigated.

With regard to passenger vehicle technology, three key findings emerged from the model results. First, recognition of the future connections between transportation and the other energy sectors is very important. Addressing climate change in a cost-effective manner requires a multi-sector perspective. Second, there is no silver bullet; given current uncertainties in future costs/efficiencies there is no clear fuel/vehicle technology winner. In many cases there was no single technology that dominated on a global scale. Third, alternative fuels are needed in response to expected dwindling oil and natural gas supply potential by the end of the century.

**Defining Critical Materials**

Graedel, T., Yale University
Originally presented on: 8/10/2009 14:35:00 - 15:00:00

Should we worry about the sustainability of supply of non-renewable resources? To study these questions, flows of metals into and from human use (the anthropogenic metal cycles) have recently been quantified for several of the common engineering metals. Still more recently, the stocks of some of those metals in active use have also been determined. Do the resource availability issues play out differently for the metals used in large quantity (e.g., copper, zinc) and those which see low-volume specialty uses (e.g., indium, dysprosium)? These issues will be discussed from the perspective of results from the Yale Stocks and Flows Project, and a recent National Research Council study on the criticality of metals.

# Relevance of Sustainability Concepts in Science and Engineering Education

## Authors:

Schrader, G. L., University of Arizona

Sustainability has become an important topic for curricular discussions at many universities. The current generation of students has been generally recognized as the most environmentally conscious group since the ecology movements of the 1960s and 70s. The challenges of climate change, diminished fresh water availability, and scarcity of material resources are also increasing expectations by industry that students become knowledgeable about environmental sustainability. The breadth of educational approaches developed at universities to address sustainability in science and engineering courses has been impressive. Common curricular elements have emerged, such as discussions of minimization of environmental footprints, reduction of resource demands, incorporation of life cycle analyses, and interactions with societal structures. At the University of Arizona, the formation of a new ?Practice School for Sustainability? seeks to further emphasize the relevancy of sustainability to industrial applications. This focused approach involves direct participation by businesses, government agencies, and campus sustainability programs to create practical projects having a comprehensive interdisciplinary perspective.

# On Sustainability Metrics for Environmental Management

## Authors:

Eason, T., U.S. Environmental Protection Agency
Garmestani, A., U.S. Environmental Protection Agency
Heberling, M., U.S. Environmental Protection Agency
Hopton, M., U.S. Environmental Protection Agency
Campbell, D., U.S. Environmental Protection Agency
Karunanithi, A. T., University of Colorado Denver
Templeton, J., U.S. Department of Transportation
White, D., U.S. Environmental Protection Agency
Zanowick, M., U.S. Environmental Protection Agency

The concept of sustainability is associated with the statement from the World Commission on Environment and Development, 1987: ?? development that meets the needs of the present without compromising the ability of future generations to meet their own needs?? Sustainability is, therefore, about the likelihood of the world persisting in a dynamic regime that is supportive of human society. But without at least semi-quantitative means of measuring progress in a meaningful manner, it is difficult to implement any sustainability strategy in any practical sense. There is, therefore, a need for scientifically sound metrics to provide quantitative measures that assess the degree to which the system is or is not on a path towards sustainability, and whether management actions taken are or are not effective in maintaining the system on a sustainable path. Inherently, sustainability questions involve a complex and interacting system that has many dimensions including ecological, economic, social, and engineering aspects. One might never fully understand such a complex system, but there is an immediate need for methods to analyze and manage it. We plan to discuss the subject of sustainability metrics generally, and then illustrate the concepts with the case of four metrics for use in environmental management: ecological footprint, emergy, green net regional product, and Fisher information. These respectively represent human burden on the environment, energy resources flow in the system, economic vitality, and overall system order and function. One should note that these metrics are meant to represent system processes and qualities that must be preserved for sustainability in a system that is dynamic, evolutionary, and, perhaps, cyclic. The illustration concludes with a discussion of the San Luis Basin Sustainability Project where we have collected data that allow us to compute these four metrics as a function of time, and to observe trends with time for about twenty years. Lastly, the outline of a management strategy on a regional scale using these concepts and metrics will be discussed.

# Sustainability as a Scientific Paradigm - An Industrial View

Sustainability plays a key role in our business processes and is important to our future success. This presentation will discuss Dow's rich history in the area, underscore the importance of vision and goals in driving results and accountability, and provide an overview of the framework Dow is using to "Set the Standard for Sustainability", one of the company's four strategic pillars. Science and technology provide a key foundation for Dow's sustainability efforts. Examples of products and processes fueled by science and technology illustrate how chemistry and engineering can foster innovation and deliver solutions to many or the world's challenges including affordable and adequate food supply, decent housing, sustainable water supplies, improved personal health and safety, as well as addressing climate change. In addition, progress on energy utilization, alternative feedstocks, and greenhouse gas emissions will be outlined. As Dow undergoes one of its most significant transformations in more than 50 years, sustainability and partnerships will play key roles.

**Practical Application of Sustainability Tools in the Chemicals Industry**

Koster, K., Cytec Industries

Originally presented on: 8/10/2009 16:35:00 - 17:00:00

Sustainability Tools developed by Cytec Industries Inc. to bring scientific analysis and sustainability closer together. Tools include:

? Use of iSUSTAIN?, an index to measure Cytec products against the 12 Principles of Green Chemistry;

? Formation of a Community of Practice to create a cross-functional forum to share, develop and implement sustainability actions

? Link between REACH and other product regulatory compliance and sustainability tools

? ?Design for SHE? initiative within Cytec

? Traditional ?footprint? initiatives

In addition Karen will provide an overview of Cytec's sustainability strategy

# Key Business Metrics That Drive Sustainability Into the Organization and Its Value Chain


## Authors:

Kearney, J. E., MedImmune, Inc.



Increasingly, business is adopting the understanding that financial performance is in fact linked to environmental and social performance; sustainability represents to business not just a lens through which to better understand and manage risks but also one for developing business opportunities and innovation. Measuring sustainability is therefore important to business if it is to more effectively manage these dimensions, as well as make informed decisions for long-term success. In seeking better ways to measure performance across the spectrum of sustainability concerns, various methodologies have emerged that attempt to better link sustainability metrics to business strategies, stakeholder issues, and resulting business objectives and actions. This keynote will highlight a few methods, including the GEMI Metrics Navigator? approach, to identifying and developing the ?critical few? metrics that link business success factors to environmental, social and financial performance, and serve as a guide to more socially responsible business practices. Examples from industry on the development and use of sustainability metrics to drive performance will also be presented.

# A Simple Sustainability Index for the Chemical Industry

## Authors:

Cohen, M., CYTEC Specialty Chemicals

A number of methodologies for providing quantification of sustainability for chemical processes have been developed over the last decade. None seem to have gained broad adoption, however, perhaps due to complexity and/or lack of availably of required information. In the current absence of full life cycle data for the bulk of raw materials used in the chemical industry, we propose a scoring system based on readily available information, and using this present a basic set of metrics, based on the Twelve Principles of Green Chemistry, that may be used to develop a sustainability index for production of a given chemical product. Applicability of this index to the social aspects of sustainability will also be considered.

# Measuring Sustainability - Tools for Sustainable Project Implementation


## Authors:

Tanzil, D., Golder Associates Inc
Griffin, B., Golder Associates Inc



In response to changing societal expectations and to gain competitive advantage, many businesses today have made public commitments to sustainability. The challenge is to translate such commitments into concrete actions at all levels of an organization ? from corporate and local operations to individual projects.

For an engineering project, implementing sustainability requires an understanding of what sustainability means and how it applies specifically to the project. Key sustainability issues are determined by various factors, including corporate strategy, local concerns, and potential positive and negative impacts from the project. A sustainability planning and performance management process developed by the Global Environmental Management Initiative (GEMI), termed the Metrics Navigator™, can be adapted to identify and prioritize issues most relevant to a project and develop appropriate performance indicators and metrics. Once a prioritized set of environmental, social, and economic indicators is developed, it can be used as input to decision-support and to monitor project performance.

This paper will provide case histories on the use of sustainability indicators to support sustainable project implementations. They include applications in environmental impact analysis (EIA), product development, and technology selection for environmental projects. Results will illustrate how sustainability performance indicators support the identification of project benefits, trade-offs, and concrete actions in implementing the sustainability commitments.

**Improving Sustainability Indices through Judicious Inclusion of Indicators**

Mayer, A. L., Michigan Technological University

Originally presented on: 8/10/2009 15:20:00 - 15:45:00

Human societies are intimately connected to their supporting environments, thus suggesting that societies should be referred to as socioecological systems. To be ?sustainable?, the conditions of socioecological systems must meet two criteria. First, the conditions must be those that can persist for the long term, given the physical constraints of the planet and other limitations. Second, the conditions must be desirable (there are undesirable system conditions that can persist indefinitely, such as violent conditions in collapsed nation-states). Desirable conditions are often described as sustainability targets or goals, and for many of these goals, one or a set of indicators are chosen to track the progress of the system towards the goals. However, the relevance of these indicators to sustainability goals, and their independence from each other, is not well-known. Often, the choice of indicators is driven solely by data availability. Some indicators, such as rates of infant mortality, have a linear relationship with sustainability; the higher the rate, the lower the sustainability of the socioecological system. Others, such as per capita GDP, have a more complicated relationship with sustainability; countries with the highest per capita GDP are not necessarily the most sustainable ecologically, socially, or even economically. Additionally, indicators often represent conditions in socioecological systems that are closely linked. These interactions may result in a situation where policies meant to improve one indicator causing a decline in another, resulting in no net increase in sustainability conditions. For example, to decrease the number of cases of illnesses caused by poor sanitary conditions, one policy may support the installation of traditional sanitation technology, which uses water to dilute and transport waste. This technology may then increase the number of people experiencing water stress by increasing water competition during drought conditions. Sustainability indices are increasingly used as a way to aggregate many environmental, social, and economic indicators into one easily-understandable value that can be tracked over time, to measure progress towards goals. If interactions among indicators are not known or accounted for, sets of correlated indicators could receive a higher weighting and therefore bias those sustainability indices that use aggregation methods such as summation. If indicators, and the aggregation of these indicators into indices, are to be useful, there must be a more thorough accounting for the ways in which they relate to sustainability goals.

# Indicator for Measuring Sustainable Product Design: a Review and Further Research


**A. Rahman Hemdi[1], Muhamad Zameri Mat Saman[2] and Safian Sharif[2]**

[1]Faculty of Mechanical Engineering, Universiti Teknologi Mara, Malaysia.
[2]Faculty of Mechanical Engineering, Universiti Teknologi Malaysia, Johor, Malaysia
Email address of corresponding author:abdulrahman643@ppinang.uitm.edu.my, zameri@fkm.utm.my, safian@fkm.utm.my



## Abstract

The focus of engineering design on achieving a superior product from functional and economical factors is now shifting towards sustainable product design which considers environmental, economical and social aspects. As the consumers have gained more awareness on the environmental issues, the demand for sustainable product has been increasing proportionally. Therefore, a new product development method is required to fulfill today's market needs. This can be achieved by integrating the sustainable elements of environment, social and economical aspects into the engineering design throughout the product's lifecycle. The life cycle assessment is being used in assisting the designer or engineer in the product design but the process possess drawback which is due to the large amount of data requirement and it is time consuming. As a result, simplified and understandable methods and tools are required in guiding the designer in developing sustainable product. This paper aims to summarize the current methods and tools for integrating sustainability perspective into the product design process. In addition, any potential room for improvement can be identified from this study. As a conclusion, this paper also proposes future research in developing a new tool for measuring sustainable product design by considering the three main elements of sustainability i.e. environmental, economical and social aspects.



Keyword : Sustainable product, product design, life cycle assessment, sustainable indicator


## 1. Introduction

Sustainable product is *a product which will give as little impact on the environment as possible during its life cycle* (Ljungberg, L. Y., 2007). Sustainable product can only be produced by taking into consideration the sustainable aspects such as environmental, technical, economical and social into the design process (Howarth, G. and Hadfield, M., 2006). Design phase plays an important role for achieving sustainability because it can be used to reduce waste, allow recyclibility and longetivity by improving a design towards sustainability. In order to achieve sustainable design, tools or indicators are needed to measure the level of product sustainability and guide the designer for developing sustainable product.

Traditionally, the life cycle assessment (LCA) can be used as a tool for evaluating product design in term of environmental aspect. Although it is a capable tool for evaluating environmental impact, it possess drawback which is due to the large amount of data requirement and it is time consuming (Khan, F.I., Sadiq, R. and Veitch, B., 2004). Therefore, simplified and understandable methods and tools are required for guiding the designer in developing sustainable product. Besides that, the sustainable perspective includes three main elements such as environmental, economical and social, require some modification of the current LCA tools. In addition, there are various sustainable attributes that can be considered as the factors in the complexity of deciding any particular design alternatives.

The research on sustainable indicator as a designer tool has been emerging rapidly. There are some developed tools which will be discussed in the next section. Meanwhile, section 3 and 4 provide some important criteria for developing the future tools and the conclusion, respectively.

## 2. Current Methods/Tools

There are available tools for measuring sustainability levels likes life cycle assessment (LCA), Ten Golden Rule, Eco Indicator, Green Pro, Life cycle index (LInX) and others. Some of the current tools are discussed further in the next section and their summary are presented in Table 1.

### 2.1. Eco Indicator 95

The Eco-indicator 95 is a quantitative distance-target based LCA methodology (Goedkoop, M., Demmers, M. and Collignon, M., 1996). In this method, a target level is set for a particular environmental effect. When the gap between the environmental impact and the target level are greater, the higher weight will be given to the seriousness of the impact. The Eco-indicator 95 is a general tool which can be used to evaluate any type of product. Besides that, it is easy to be applied by a designer without having deep knowledge on the environmental issues. However, the evaluation does not include the economical factor such as cost, resources depletion and technology.

### 2.2. Eco Indicator 99

Eco-indicator 99 is a modification of Eco-indicator 95 which is based on damage oriented method for life cycle assessment. It was developed based on the three main damage categories which include towards human health, ecosystem and mineral resources (Goedkoop, M. and Spriensma, R., 2001). The Eco-indicator 99 is a generalized tool which can be used to evaluate any products. Besides that, it is accepted as international standard and well documented (Li, Jianzhi., Wu, Zhenhua and Zhang, Hong-Chao.,2008). However, limitation of this tool, the assessment does not include the analysis of cost and technology.

### 2.3 Life Cycle Index, (LiNX)

Khan, F.I., Sadiq, R. and Veitch, B. (2004), had proposed an indexing system for evaluating process design. The sustainable attributes were grouped into houses and colonies such as environmental colonies, cost house and technology house. This generalized tool can be used for screening and evaluation of any type of products and process design. The application of LiNX can be broaden to the product design for evaluating the sustainability level (Khan, F.I., Sadiq, R. and Veitch, B., 2004). However, the analysis boundary is limited to the "cradle to gate" which does not cover the use and end of life phase of product.

### 2.4 Green Pro

Green Pro was developed by Khan, F.I., Natrajan, B.R. and Revathi, P. 2001, a systematic methodology for process design that consider assessment and minimization of environmental impact. The analysis includes environmental, technology and economical factors at the design stage to determine cost-effective solution. The guidance for decision making by applying Multi Criteria Decision Making (MCDM) analysis is one of the gain of this tool. But, the assessment is still lacking on the sustainable perspective by missing the analysis of social aspect. Besides that, the boundary analysis is only limited to "cradle to gate".

### 2.5 Green Pro I

A year later, the improvement of Green Pro was then developed by Khan, F.I, Sadiq, R. and Husain, T., 2002 by introducing Green Pro I. The modification concerns on two areas which are broadening the boundary analysis to the "cradle to grave" and applying Fuzzy Multi Criteria Decision Making for better decision making analysis. However, it still overlooks the analysis of social aspect.

### 2.6 Ten Golden Rule

Ten Golden Rule is a qualitative analysis method in providing common foundation which can be used as a basis and guidelines for development of specific product design (Luttropp, C and Lagerstedt, J., 2006). The rules can be customized depending on the specific requirement of a product. However, the designer or user must have background knowledge to be able to make sound use of these 10 rules. On the other hand, the analysis result may differ, depending on the user knowledge and experience.

### 2.7 Simplified LCA and Environmental Responsible Product Assessment (ERPA)

Simplified LCA (SLCA) is a new method which provides less cost, time and effort but having similar result of full LCA (Hur, T., Lee, J., Ryu, J., Kwon, E., 2005). SLCA was developed by reducing the scope or data needed using available secondary data or generic databases instead of trying to collect primary data. According to Hur, T., Lee, J., Ryu, J., Kwon, E. (2005), the actual site data should be used for manufacturing and usage stage while the secondary data can be used for material production, component manufacturing and material disposal processes. Meanwhile, the Environmentally Responsible Product Assessment (ERPA) matrix was used to present information on product environmental aspect in systematic manner. Although SLCA applied "cradle to grave" boundary, it does not cover the analysis of other two sustainable elements such as social and economical aspects.

### 2.8 Volvo Black, Grey and White Lists

The Volvo Black, Grey and White Lists is a qualitative approach used to guide the designer or engineer on selection of materials or substances in Volvo's production process. It lists the chemical substances which must not be used known as black list, should be limited in use referred as the grey list and the white list substance which may critical from the health and environment point of view (Byggeth, S. and Hochschorner, E. 2006).

Table 1. Summary of several tools for measuring sustainability

| No. | Name of Indicator | Description |
|---|---|---|
| 1 | Eco Indicator 95 | • can be used by designer who has a little knowledge on environmental impact<br>• for comparison between alternative design<br>• it is a general tool which can be used for any type of product<br>• does not include cost, material depletion and technology |
| 2. | Eco Indicator 99 | • include damage of resources depletion, human health and ecosystem<br>• follows ISO14042 of life cycle assessment<br>• can be used by designer who has a little knowledge on environmental impact<br>• a general tool and can be used for any type of product<br>• does not include cost and technology |
| 3. | Green Pro | • used to evaluate process design alternative by considering environmental and economical aspect<br>• apply cradle to gate approach<br>• apply MDCM (Multi Criteria Decision Making) to help and guide the decision makers in identifying the most desired solution<br>• does not include use, disposal, recycling, reuse and remanufacture<br>• lack of sustainable perspective (does not include social aspect in sustainable evaluation) |
| 4. | Green Pro I | • improvement from Green Pro<br>• apply cradle to grave approach<br>• evaluation is based on environmental risk, cost and technical feasibility<br>• apply fuzzy based approach into Multi Criteria Decision Making (MDCM) to evaluate some subjective attributes<br>• lack of sustainable perspective (does not include social aspect in sustainable evaluation) |
| 5. | Life Cycle Index (LInX) | • introduce indexing system which is very flexible and can be changed to the requirement of the user depending on the scope/focus of study<br>• aids in short listing or highlighting the areas which require detail study<br>• used as a rapid screening tool for LCA based decision-making<br>• require less time requirement than conventional LCA<br>• evaluate a product based on 4 attributes (cost, politic, environment, technology)<br>• does not include use, disposal, recycling, reuse and remanufacture |
| 6. | Ten Golden Rules | • generalized set of guidelines (can be used for any product)<br>• the rule provide generic guidelines, it has to be customized according to specific product.<br>• can be used by designer to develop the personal version from generic guidelines<br>• analysis of result may differ for different designer because it depends on user's knowledge |

| | | |
|---|---|---|
| 7. | Simplified LCA (SLCA) & Environmentally Responsible Product Assessment (ERPA) | • SLCA generate more information that can be used for new design when developing new product<br>• provide less cost, time and effort but having similar result with full LCA<br>• data needs were reduced by using available secondary data/generic databases instead of trying to collect primary data<br>• ERPA matrix can be used to present information about product environmental aspect in systematic manner and to identify area for improvement<br>• does not include analysis on economic value/cost and impact to human health |
| 8. | Volvo's Black, Grey and White Lists | • list of chemical substances which:<br>o must not be used (black list)<br>o should be limited in use/chemical substances which may be critical from health & environmental point of view (grey list)<br>o potentially less hazardous (white list)<br>• apply in Volvo's production processes |

## 3. Developing Future Tools

Based on the literature and analysis of the current tools, consideration of some important criteria is essential in developing future tools. The proposed tool should follow the ISO 14040 requirement in order to be accepted internationally. Besides that, it must include all the sustainable elements, cost evaluation and provide guidance for decision making and also adaptable to the designer's works.

### 3.1. International standard ISO14040

The indicator of sustainable product design can be developed by assessing each life cycle of the product. LCA methods can be the basic guidance for designing an indicator. The implementation of LCA is standardized according to the Environmental Management-Life Cycle Assessment-Principles and Framework (ISO 14040, 2006). According to this standard, the development of LCA should consist of four stages such as:

- *The goal and scope of definition phase*
  The goal usually states the intended application, reason and audience while the scope usually includes product system to be studied, system boundary, data requirement, assumptions and limitations.
- *The inventory analysis phase (Life Cycle Inventory, LCI)*
  LCI involves collection and calculation procedures to quantify relevant inputs and output of a product system. The data collection may in this phase consists of energy and raw material input, product and waste, emission to air, discharge to water, and other environmental aspect.
- *The impact assessment phase (Life Cycle Impact Assessment, LCIA)*
  The impact assessment phase is intended at evaluate the significance of potential environmental impacts by using the LCI results. The flow of LCIA is shown in Figure 1 which has mandatory and optional elements. LCIA starts from the selection of impact categories, category indicators and characterization models.

Other optional elements such as normalization and weighting can be applied depending on the goal, scope and corporate priorities. For evaluating the LCIA, the use of fuzzy theory and artificial neural network is very helpful in reducing the error caused by data uncertainties (Zhou C-C et al. 2008; Benetto, E., Dujet, C., Rousseaux, P. 2008; Tan, R.R., Culaba, A.B. and Aviso, K.B. 2008; Khan, F.I, Sadiq, R. and Husain, T. 2002).

- *The interpretation phase*
  During the interpretation phase, both the findings from LCI and LCIA are considered. The interpretation phase should deliver results that are consistent with the defined goal and scope. In addition, it should also highlight conclusions, explain limitations and provide recommendations. Another important thing is the interpretation must be readily understood, complete and consistent with presentation of the LCA result.

Another essential element of LCA is the report and critical review. The LCA result and conclusion should be presented in adequate form to the intended audience, addresses the data, methods and assumptions together with the limitations applied in the study. Meanwhile, the critical review is essential to verify whether LCA has met the requirements for methodology, data interpretation and reporting.

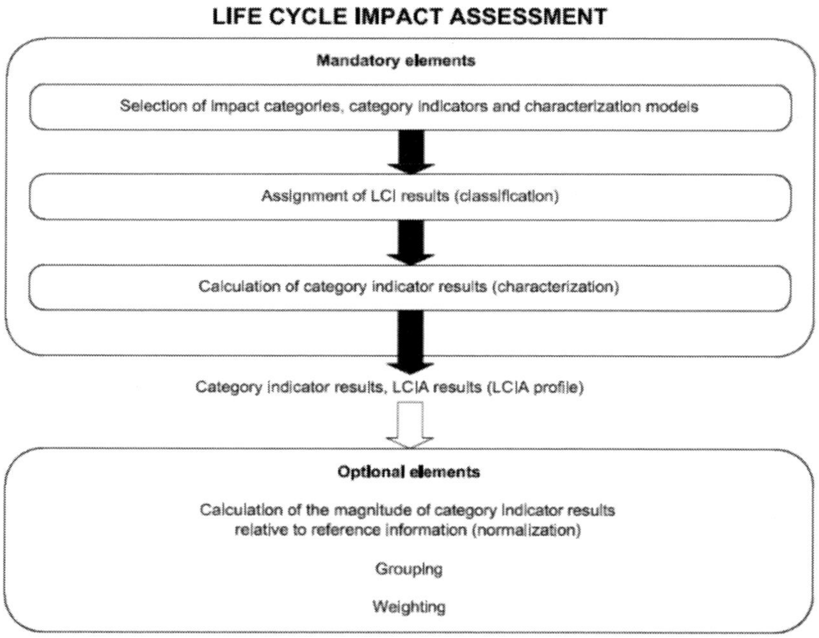


**Figure 1:** Mandatory and optional elements of LCIA phase (ISO 14040, 2006).

## 3.2. Sustainable Perspective

For developing a future tool, the three main sustainable elements must be integrated into the evaluation tools. The environmental, economical and social aspects have to be set as a core principle in designing the indicator for product sustainability.

Each of the three sustainable elements has an important role in maintaining our world's sustainability. Based on that, the future tools should guide the designer or engineer to design products which have the lowest environmental impact, low material usage, zero waste, lowest cost and lowest effects toward human health and nuisance.

In order to evaluate the sustainability of product design, the sustainable elements are then separated into several sub elements or sustainable attributes. The proposed sustainable attributes are shown in Figure 2. The sustainability level of product design alternatives are evaluated based on each attribute throughout its life cycle.

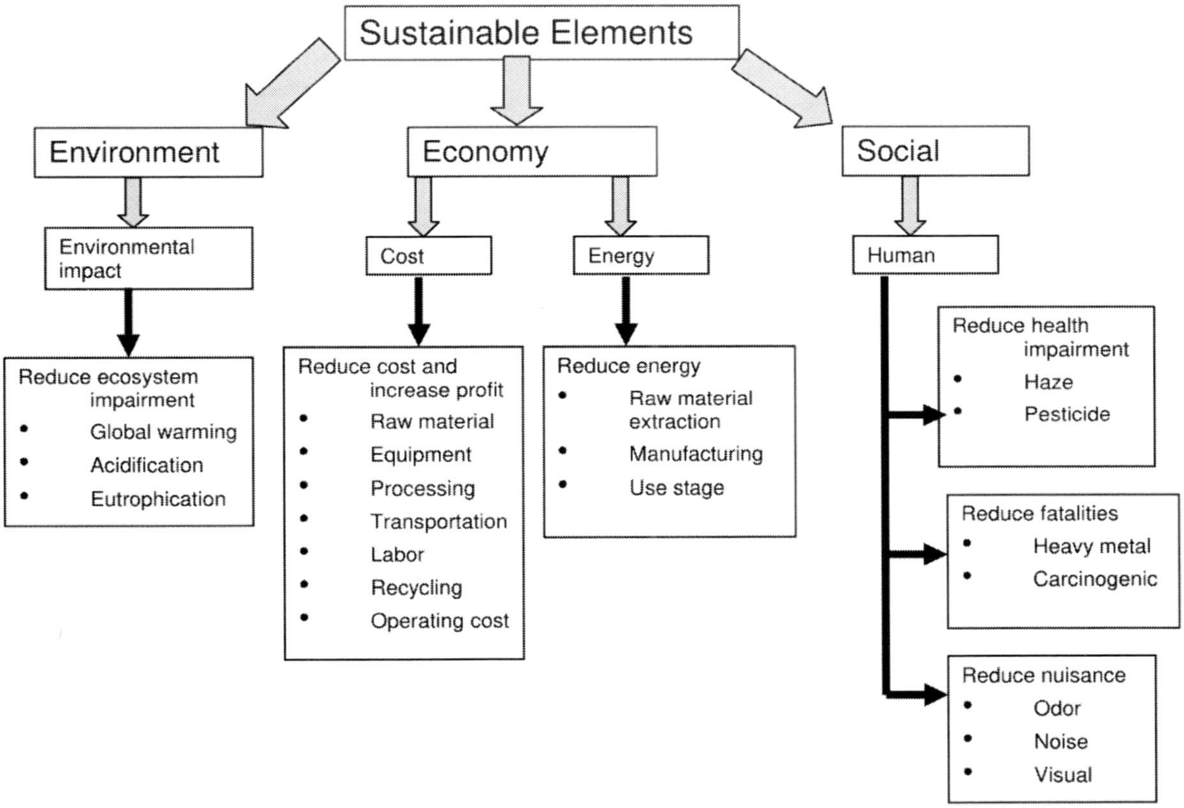


**Figure 2:** Proposed sustainable attributes

### 3.3. Cost Evaluation

Cost evaluation is one of the most important criteria that must be included into future tools. Instead of comparing the design alternative with respect to the environmental and social impact, a good tool should also be able to distinguish each alternative and design changes in term of economic factors. The economic factors can be in term of cost such as cost of producing a product, total life cycle cost of product and eco-costs. The cost of producing a product includes research and development costs, material costs, manufacturing costs and operational costs. Meanwhile, the total life cycle cost is the summation of all costs associated throughout the life cycle starting from research and development costs, manufacturing costs, operational costs, support or maintenance costs until the retirement and disposal costs. Whereas eco-costs are costs of effluent waste treatment, costs of effluent waste control and disposal, costs of environmental management systems, costs of eco-taxes, costs of rehabilitation, costs of energy and costs saved from recycling and reused strategies (Senthil, Kumaran D., Ong, S.K., Nee, A.Y.C. and Tan, Reginald B.H., 2003)

By evaluating and comparing each design alternative with costs analysis and then relating it with the environmental and social factors, the result provides a good basis for decision making of alternatives. The result will give a clear and strong judgment to the decision maker for deciding the best solution. Besides that, studies conducted by Gehin, A., Zwolinski, P. and Brissaud, D. (2008) found out that the major weakness of some tools is that they do not include economical value in their judgments. It was assumed that creating environmentally friendly product is profitable, but there are either no or weak justification given to support the judgment. The decision makers are usually economist and some of them are not environmentally knowledgeable. They only want to know "WHY rather than HOW" to produce sustainable product (Gehin, A., Zwolinski, P. and Brissaud, D., 2008).

### 3.4. Decision Making

Selection of the best design among the various alternatives is a very important task in the product development stage. Decision making is a complex process where it involves many criteria to be considered and also must fulfills the objectives or goals of an organization. For example, in order to select the best sustainable product design, the criteria of environmental, economic and social aspects have to be considered simultaneously. Thus, it will increase the complexity of decision making process.

Based on that requirement, the future tools should have the ability to guide the designer or engineer in decision making. It can be done by integrating the decision making process into the tools. Applying the Multi Criteria Decision Making (MCDM) process can be part of the solution. Sadiq, R. and Khan, F. I. (2006), proposed the probability theory and fuzzy set theory to guide decision making by converting qualitative items into fuzzy number, grouping the attributes and ranking the alternatives. Apart from providing guidance in decision making, it will benefit the designer or engineer in giving a better visualization of the sustainability condition of the design alternative and thus, provide strong justification for each decision.

### 3.5 Daily Work

Another important aspect which needs to be considered in developing a future sustainable product design indicator is to make it adaptable to the designer or engineer daily work (Gehin, A., Zwolinski, P. and Brissaud, D., 2008). It is not applicable for producing a tool which is claimed to cover all the sustainable attributes and using artificial intelligent software if the designer need to study or taking environmental course in order to sound use of the tools. The future tools must be easy to use and able to produce easy interpretable results, so that it can be adapted into designer's daily works. Quantitatively evaluation of sustainable attributes and presenting the analysis result by using graphical method, followed by ranking the design alternatives using a table can be the best option. In the case of design improvement, instead of measuring the sustainability level and finding a weak point, the tool should be able to propose or provide some suggestions on how to assist the designer in improve the existing design.

## 4. Conclusion

There are some improvements that can be done in developing a sustainable product design indicator. Although some of the indicator follows the international standard ISO 14040, but they are still lacking in terms of sustainable perspectives. The three elements of sustainable perspective such as social, economical and environmental are often found missing in the tools. Therefore further works should focus in developing an indicator which can evaluate sustainability by focusing on all the three sustainable elements and by considering the total product life cycle starting from material extraction to the product end of life.

Besides that, the cost evaluation by considering the total cost of product life cycle and eco-costs has to be integrated into the tool. It can be used as a judgment for decision making and showing that the sustainable product is not only reducing environmental impact but also increases the company's profit.

In addition, the new indicator should be more comprehensive, easy to use by the designer or engineer and can be adapted into their daily works. This is possibly achievable by considering qualitative together with quantitative approaches in the sustainable evaluation. Other than that, the result of evaluation should be presented in interpretable manner in order to assist designer or engineer in decision making process. The element of artificial intelligent theory such as fuzzy and neural network can also be integrated into the evaluation and decision making process.

## References


Benetto, E., Dujet, C., Rousseaux, P. (2008). *Integrating Fuzzy Multicriteria Analysis and Uncertainty Evaluation in Life Cycle Assessment,* Environmental Modelling & Software, 23: 1461-1467

Byggeth, S. and Hochschorner, E.(2006), *Handling trade-offs in Ecodesign tools for sustainable product development and procurement.* Journal of Cleaner Production, Vol.14, pp.1420-1430

Gehin, A., Zwolinski, P. and Brissaud, D.(2008), *A tool to implement sustainable end-of-life strategies in the product development phase.* Journal of Cleaner Production, 16, pp. 566-576

Goedkoop, M. and Spriensma, R. (2001), *Eco Indicator 99 A Damage Oriented Method for Life Cycle Impact Assessment : Methodology Report,* Product Ecology Consultants (PRé), 3rd ed. Netherlands

Goedkoop, M., Demmers, M. and Collignon, M. (1996), Eco *Indicator 95: Manual for Designer,* Product Ecology Consultants (PRé), Netherlands.

Howarth, G. and Hadfield, M. (2006), *A Sustainable Product Design Model.* Material and Design, Vol. 27, pp. 1128-1133

Hur, T., Lee, J., Ryu, J., Kwon, E.(2005), *Simplified LCA and matrix methods in identifying the environmental aspects of a product system,* Journal of Environmental Management, 75, pp. 229-237

ISO 14040 (2006), *Environmental Management-Life Cycle Assessment-Principle and Framework.,* International Standardization Organization.

Khan, F.I, Sadiq, R. and Husain, T.(2002), *GreenPro-I: a risk-based life cycle assessment and decision-making methodology for process plant design.* Environmental Modelling & Software, 17, pp. 669-692

Khan, F.I., Sadiq, R. and Veitch, B.(2004), *Life Cycle iNdeX (LInX): a new indexing procedure for process and product design and decision-making*, Journal of Cleaner Production, 12, pp. 59-76

Li, Jianzhi., Wu, Zhenhua and Zhang, Hong-Chao.(2008), *Application of Neural Network on Environmental Impact Assessments Tools.* International Journal Sustainable Manufacturing, Vol. 1, Nos. 1/2. pp 100-121

Ljungberg, Lennart Y.(2007) *Materials selection and design for development of sustainable products.* Materials and Design, 28, pp. 466-479

Luttropp, C and Lagerstedt, J.(2006), *EcoDesign and The Ten Golden Rules : generic advice for merging environmental aspects into product development*, Journal of Cleaner Production, Vol. 14, pp 1396 – 1408

Sadiq,R. and Khan, F. I.(2006), *An integrated approach for risk-based life cycle assessment and multi-criteria decision making – selection, design and evaluation of cleaner and greener processes.* Business Process Management Journal, Vol. 12 No. 6, pp. 770-792

Senthil, Kumaran D., Ong, S.K., Nee, A.Y.C. and Tan, Reginald B.H.(2003), *A Proposed Tool to Integrate Environmental and Economical Assessments of Products.* Environmental Impact Assessment Review, 23, 51-72.

Tan, R.R., Culaba, A.B. and Aviso, K.B. (2008). *A Fuzzy Linear Programming Extension of the General Matrix-based Life Cycle Model*, Journal of Cleaner Production, 16: 1358-1367

Zhou C-C et al. (2008). *Multi-objective Optimization of Material Selection for Sustainable Products: Atificial Neural Networks and Genetic Algorithm Approach*, Journal Material and Design, doi:10.1016/j.matdes.2008.06.006

**Pathway to Sustainable Energy**

Arastoopour, H., Illinois Institute of Technology
Originally presented on: 8/10/2009 16:10:00 - 16:35:00

The Energy and Sustainability paradigm embodies all of the services required to continue to improve human, social, and economic well-being while preserving resources and the environment for future generations. Specifically, our approach to sustainable energy implements a least-cost strategy to continue the decarbonization of the global energy system, conserve resources and energy through material recycling and energy efficiency enhancement, increase power reliability and security, and minimize pollution.

Sustainable energy connects three major issues?economy, environment, and security. A comprehensive national and international energy and sustainability policy and matrix are needed to achieve the desired emerging paradigm including: zero-emissions plants and vehicles, ?perfect? electricity, sustainable built environment, and full-life-cycle accountability that considers trends toward continued growth in population and energy consumption per capita, continued depletion of oil and gas resources, and the increasingly significant role of environmental issues.

This presentation will discuss short- and long-term issues in the development and implementation of the pathway to sustainable energy paradigm, the role of the distributed renewable energy system in the sustainable built environment, the connection of the stationary and transportation energy distribution systems (e.g., plug-in hybrid vehicles), and the smart-grid concept.

# Measuring Sustainability in Houston, Texas

A project has been initiated at Rice University to measure sustainability using the City of Houston as a case study. The goal of the project is to find objective methods of measuring sustainability, combining experts from the natural sciences, environmental law, sociology and economics. This presentation will focus upon the initial discussions and decisions regarding the issues of primary concern for measurement. Of particular interest are issues such as the equitable distribution of jobs, whether those most at environmental risk are benefitting from such risk, economic structure for sustainability, environmental limits, the flow of materials and energy through the City, the health and fate of ecological systems, food security and cost, full cost pricing of services, and other similar issues.

**A Fuzzy-Logic-Based Triple-A Template for Industrial Sustainability Enhancement**

Liu, Z., Wayne State University
Huang, Y., Wayne State University

Industrial sustainability is pursued to achieve the long-term sustainable development (SD) of a given industrial system that could be defined as a network of industrial sectors, each composed by a number of entities. Industrial sustainability problems are always difficult to be thoroughly investigated and further optimized, because of the problem size and scope that carry highly complexity and inherent uncertainty that are associated with data, information, and knowledge. Therefore, current industrial practices on industrial sustainability enhancement are mostly scenario based, where the enhancement strategies are heuristically generated and improved until the degree of sustainability of the system has been increased based on the criteria selected by the involved parties.

One of the major challenges in industrial sustainability research is how to effectively handle uncertainty. In this paper, a fuzzy-logic-based Triple-A Template that can be used to conduct more effective sustainability studies in three consecutive steps: Assessment, Analysis, and Action. First, a fuzzy-logic-based multilayer sustainability assessment is applied to a given industrial zone system problem. With carefully defined fuzzy sets for the triple bottom lines (i.e., economic, environmental, and social sustainability), the sustainability status of the zone can be appropriate determined under uncertainties of various types. The assessment gives the classified sustainability status of not only at the industrial zone level, but also its subsystems. Then, the industrial zone's sustainability problem is analyzed by the approach of design of experiments in order to determine systematically the causes of those poor aspects of sustainability and the relationships among them. In the last step, potential actions will be proposed on the identified causes. Instead of proposing limited numbers of strategies heuristically and comparing their associated scenarios, general action strategies are taken in the proposed methodology and fuzzy logic based optimization is further applied based on the relationships derived before to obtain enhancement strategies under given constraints and uncertainties.

The main advantage of the introduced methodology is its capability of effectively and systematically identifying the optimal enhancement strategies for a given industrial system problem under uncertainties. The methodological efficacy will be illustrated through analyzing the sustainability issues and developing strategies for enhancing the sustainability of an automotive manufacturing centered industrial zone.

**Reduction of Atmospheric CARBON DIOXIDE with A Modified Solar Updraft Tower**

Moolick, R. T., Arizona State University

The major challenges of carbon sequestration is the extraction of significant amounts of carbon dioxide and the production of a stable carbon storage configuration while achieving a negative net carbon footprint. This can be achieved through a modification of the solar updraft tower generation facility such as the one planned for construction in Australia. The modified solar updraft tower uses both biological and chemical processes to maximize carbon dioxide removal. This integrated system immobilizes carbon in two forms. The biological process stores carbon as a solid organic carbon (biomass) form. The chemical process stores carbon as a solid, stable carbonate. The combined processes will result in a greater carbon removal than either process separately. These processes are powered by a small fraction of the energy generated by the solar updraft tower. Ultimately the goal of removing and storing carbon dioxide from the atmosphere with a net negative carbon footprint is achieved.

**Establishing Biofuel Metrics**

Monk, J., University of California, San Diego
Dang, D., American Institute of Chemical Engineers

Biofuels have the potential to provide a clean, sustainable and affordable supply of renewable transportation fuel. In order for this potential to be realized, candidate feedstocks, farming practices and technologies must be identified. The purpose of our project is to establish a set of metrics that can be used to quantifiably evaluate biofuels. Presently there is no widely accepted framework for analyzing different biofuels. Our goal is to establish a transparent, credible and comprehensive set of biofuel metrics. We plan to establish metrics for: (1) Cost per ?standard? mile (psm), (2)Net renewable energy per acre. Corrected for differences in engine efficiency, (3)Potable water use (psm), (4) Environmental impact on water quality (psm) (5) Environmental impact on air quality (psm) (6) Food production displaced (psm)(7)Adverse impacts on bio-diversity resulting from changing use of any land or water (psm)(8) Percent of sustainably available feedstock required for 10% of the U.S. demand. The metrics focus on the entire chain of use including growth, harvest, transport, conversion and distribution. We are presently focusing on establishing metrics for 4 fuel types: Soybeans to biodiesel, Biomass to biodiesel, Corn to ethanol and cellulose to ethanol. The process of establishing these metrics will lead to a better understanding of current knowledge gaps. We will present our metrics in a transparent format to highlight key assumptions, knowledge gaps and areas for improvement.

**Chemically Untreated Micro-Fibrillated Cellulose as a Replacement for Petrochemical Derived Plastic Films**

Blodgett, S., 1411 North Monroe Blvd

Cellulose, the main ingredient of plant cell walls, is the most abundant natural resource on the planet. It is a renewable, non-toxic, bio-degradable material. The main objective of this experiment was to determine whether chemically untreated cellulose can be used as a replacement for petrochemical derived plastic films. Micro-Fibrillated Cellulose was produced by extensive mechanical beating of commercial bleached softwood Kraft pulp. This highly fibrillated pulp was diluted to form a cellulose and water suspension of about 0.2% consistency. The cellulose was then formed into a film by a vacuum forming method. The films were dried under varying temperature and pressure in a heat press. The resulting films were characterized for their mechanical properties, water absorbent properties, and barrier properties.

**Social, Economic and Environmental Metrics for the Sustainable OPTIMIZATION of Chemical and Petroleum Processes**

Shadiya, O. O., Oklahoma State University
High, K., Oklahoma State University

This research is focused on investigating sustainability metrics for the optimization of chemical and petroleum processes. Traditionally, engineers designed processes to meet economic goals; however with the awareness of sustainability, engineers are now considering other constraints such as resource usage, environmental impacts, social benefits and economics. Several metrics and indicators for quantifying sustainability exist; yet, there is no straight forward computer aided tool and path that engineers can follow in order to design for sustainability. Investigating and validating available indexes and metrics and implementing the three major sustainability criteria into process design and optimization are the goals of this research. A well defined methodology that will include process optimization under social, economic and environmental constraints is being developed. This involves building upon existing multi-objective algorithms that handle selected sustainability criteria and constraints.

This research will contribute to sustainability development in chemical engineering as benefits such as reduced pollution minimize resource usage and more economic products will be thoroughly explored. Also, this tool will assist in reducing the negative impact to the ecosystem as this methodology will be applicable other processes industries such as chemical, petroleum and food. Therefore it will benefit the society in the long run through the development of more economic and environmentally viable processes.

This poster will introduce our methodology as well as present results from evaluating the Green Metric system and our waste minimization procedure. The waste minimization procedure involves a simultaneous approach to maximize profit, while minimizing waste through source reduction with the implementation of multi objective optimization with process simulation. To accomplish this objective, the acrylonitrile process was studied. The approach involved four steps: process modeling and analysis, identification of process alternatives, selection of process alternatives and incorporation of multiobjective optimization. The simulator ASPEN PLUS? version 12.1 was used to model and evaluate the process. Process alternatives were identified from both literature and sensitivity analysis. Multiobjective optimization helps initially in quickly eliminating inferior alternatives and then in enhancing the superior alternative to develop an efficient process. Through this work, a general methodology was developed for the application of process optimization to process design. Using the concepts of multiobjective optimization, a sensitivity curve that shows the effect of waste treatment on the savings for the process can be developed. Based on the sensitivity analysis results, the Plug Flow reactor scheme was the better alternative. After process modification, the process can earn a maximum net savings of $ 2.64 x 107 in seven years, an increase of 69% over the base case. At the other end of the spectrum, the modified process could generate waste of 4153 lb/hr, a 38% reduction compared to the base case.

**Coordinated Sustainable Product and Supply Chain Design and Modeling**

Metta, H., University of Kentucky
Badurdeen, F., University of Kentucky
Goldsby, T., University of Kentucky

Abstract: Sustainability in business operations calls for making the products, processes and the systems, i.e. the entire supply chain, more sustainable. Therefore, to achieve continued profitability while ensuring environmental protection and societal well-being, it is imperative that product design and process design be integrated and, simultaneously, product design and supply chain design be coordinated. The former, in the way of concurrent engineering, has received considerable attention in recent times. However, the design of effective sustainable supply chains (SSC's), from a total sustainability perspective, where the strategic optimization of supply chain design is to be integrated with sustainable product design is still lacking. This is partly due to businesses not recognizing the significance of coordinated product and supply chain design (for sustainability) and also because of the sheer daunting complexity of the modeling task. Contributing significantly to this complexity are such factors as uncertainties in timing, quality and quantity of returns, stochastic routings of reverse flows for closed-loop SSC's using reverse logistics and the problematic task of quantifying product, process and SC impact on difficult-to-quantify areas like environmental and societal wellbeing. This paper will present an integrated approach to the quantitative modeling and evaluation of coordinated sustainable product and supply chain design considering the 6R-based (reduce, reuse, recycle, recover, redesign and remanufacture) approach to closed-loop material flow.

**A Novel Process for Biological Nitrogen Removal From Dairy Wastewater Using Constructed Wetlands**

Huchzermeier, M., State University of New York College of Environmental Science and Forestry
Tao, W., State University of New York College of Environmental Science and Forestry
Wen, J., State University of New York College of Environmental Science and Forestry

Dairy farms generate liquid manure, milking washwater, and contaminated runoff, which are often characterized as ammonia-rich. Dairy wastewater, left untreated, poses environmental effects and health concerns when discharged to the environment. Traditional nitrogen removal systems require an energy intensive two-step process: nitrification-denitrification. In our integrative constructed wetland treatment system, total nitrogen (TN) removal is achieved by a nitritation-anammox process. Partial nitrification (nitritation) is responsible for producing nitrite, a reactant for the anammox process. Autotrophic anammox bacteria provide a short-cut pathway for TN removal while requiring less energy, producing less waste, and minimizing greenhouse gas emissions. In the present study, four subsurface flow constructed wetlands have been operated in series with three free water surface constructed wetlands since January 2009. All of the constructed wetlands were seeded with anammox bacteria. Two subsurface flow wetlands were packed with marble chips and the other two with pea pebbles. The free water surface wetlands contain silt loam as rooting substrate. Aquatic plants present in the free water surface wetlands include cattails, bulrushes, reeds, and pondweed. The system was operated in batch mode with synthetic dairy wastewater (targeted ammonia-N concentration at 90 mg/L) from January to April 2009. Each wetland had a hydraulic retention time of 7 days and a surface area of 0.21 square meters. The marble chip subsurface flow wetlands (BM) were able to achieve a TN mass removal rate of 87 mg/d. The subsurface flow wetlands using pea pebbles (BP) had TN removal rate less than 38 mg/d and showed a decreasing trend, likely due to the absence of nitritation at acidic conditions. In series with a free water surface wetland (CW1), the marble chip series (BM-CW1) achieved a TN removal rate of 347 mg/d. In series with another free water surface wetland (CW2), the pea pebble series (BP-CW2) achieved a TN removal rate of 229 mg/d. The BM-CW1 series achieved up to 100% ammonia removal efficency and 69% TN removal efficiency. An additional free water surface polishing wetland (CWp), which was in series with BM-CW1 and BP-CW2 series, was able to further remove TN up to 179 mg/d. Since May 4, 2009, the system has been tested to treat more ammonia-rich dairy wastewater (ammonia-N concentration at 200 mg/L). Increasing regulation for dairy farms to manage waste streams will require that farmers implement wastewater management plans. Constructed wetlands provide a low-cost, low-energy on-site treatment alternative for effective biological nitrogen removal.

**Useful Metrics for Evaluating Energy Legislation in U.S. Congress**

Tabone, M., University of Pittsburgh

Recent developments in energy prices as well as in attitudes towards climate change have sparked new types of energy legislation.

Specifically, the 2007 Energy Independence and Security Act included the Renewable Fuel Standard, a mandate for increased production of ethanol and other advanced renewable fuels over the next 15 years.

Congress must be careful in its pursuit to speed up progress. New studies have called into question the environmental and economic benefits of renewable fuels and Congress has considered removing the mandate in subsequent legislation. A steady and consistent energy policy is needed to spur investment and development and this consistency is not well served by multiple large course corrections in response to unforeseen consequences. In order to create a consistent energy policy, congress needs to be presented with comprehendible and consistent information. Through an assessment of metrics used during the writing and creation of the renewable fuel standard, this paper advocates and lays groundwork for the creation of a standard system of metrics for evaluating energy legislation. This system should not only serve to well inform Congress of the multiple environmental, economic, and social impacts of a proposed piece of legislation, but also be able to resolve and explain conflicts between metrics presented by various government agencies and interest groups. Specifically, this study:

· Presents metrics used during congressional proceedings regarding the creation and passing of the 2007 Renewable Fuel Standard,

· Presents criticisms of these metrics,

· Presents conflicts between the metrics,

· Attempts to resolve conflicts,

· Recommends methods for presenting a standard system of metrics,

· Recommends a standard system of metrics for evaluating renewable fuels.

This paper is the result of an internship with AIChE through the Washington Internships for Students in Engineering (WISE) Program.

**Life-Cycle Design: Using LCA in Integrated Product and Material Design**

Tabone, M., University of Pittsburgh
Cregg, J., University of Pennsylvania

This study will use Life Cycle Assessment methods to create and demonstrate a more effective framework for designing sustainable products. Sustainable or ?Green' products have become increasingly prevalent over the past 10-20 years as evidenced by the growth in bio-based materials, green labeling initiatives, etc. This shift represents an increase in consumer concern for the environment as well as an increase in corporate concern for pre-adopting sustainable practices which will soon be necessary. Unfortunately there is no recognized standard system for evaluating the sustainability of a product and thus many environmental marketing claims are left unjustified to the consumer. Additionally, few systems exist for evaluating materials and decisions during the design of sustainable products. As a result, green products are typically re-designed products with concern to one metric that represents sustainability, e.g. the exclusion of one or two widely known contaminants (e.g., PVC, BPA). These designs reflect a lack of communication between material and product designers regarding sustainability as well as a lack of attention to the unintended environmental impacts of a product. Given the ability to comprehensively evaluate the environmental impacts of synthetic material production as well as product construction and use through Life Cycle Assessment, it seems there should be an effective method for designing sustainable products while accounting for multiple categories of environmental impacts. Through case studies of recently designed polycarbonate product alternatives as well as PET product alternatives, this study will assess strengths and short-comings of current sustainable chemical product design methods and offer alternatives for more adequate environmental assessment during the design process.

**Development of Sustainable Integrated Aquaculture Systems with Assessment of Environmental, Social, and Economic Implications**

VanderLugt, K. R., University of Arizona
Fitzsimmons, K., University of Arizona

Current methods of large-scale aquaculture systems are not sustainable, and pose a threat not only to wild fish stocks but also to the environment. Intensive production of shrimp, for example, has been linked to the degradation of the environment as a result of habitat modification and/or effluent discharge. Although aquaculture effluent is regulated under the Clean Water Act, Section 104, 33 U.S.C., even the discharge of treated aquaculture effluent into open waters may impose negative environmental impacts in the long term. To minimize the negative impacts on the environment innovative methods of aquaculture are now being explored to promote sustainable aquaculture production. One such method is the re-utilization of aquaculture effluent in value-added commercial applications.

Untreated aquaculture effluent contains high levels of nitrogenous (i.e. ammonia and nitrates), phosphorous, and potassium nutrients resulting from uneaten fish feed and fecal material. Nutrient concentrations are also proportional to the quantity of fish in the production system. Hence, the negative environmental impacts resulting from the discharge of aquaculture effluent from intensive systems will continue to intensify without the adaptation of sustainable practices. As the aquaculture industry continues to expand globally, cost-effective sustainable methods of water quality management must be implemented both in the United States and internationally.

To address the management of nutrient laden aquaculture effluent, an experiment was performed at the Environmental Research Laboratory, in Tucson, Arizona to evaluate the potential environmental and economic benefits of using aquaculture effluent from the fish Tilapia (*Oreochromis niloticus*) to grow several food crops including Barley (*Hordeum vulgarein*), Basil (*Ocimum basilicum*), and soybean (*Glycine max*), in a soil based medium. Re-circulating integrated aquaculture-agriculture (RIAA) systems were established by irrigating individually potted plants with effluent from 5000 L pools using an above surface drip system. Integrated systems combine aquaculture and agriculture practices in which the nutrient rich aquaculture effluent is used to irrigate plants. This is particularly advantageous in terms of nutrient utilization as nitrogenous wastes contained in the aquaculture effluent are utilized as a form of fertilizer by the plants.

Several successive experiments were run in which a different food crop was grown in each experiment each lasting between 10 and 14 weeks. In each experiment, three RIAA systems referred to as Trts. 1, 2, and 3 were simultaneously run with densities of Tilapia at 3, 6, and 9 kg $m^{-3}$ respectively. Water from each 5000 L pool was distributed to a total of 300 plants divided equally between one and ½ gallon pots filled with 1:3 soil (Sunshine mix #1 ) to sand mixture. Pots were set up in a common pot array and placed in irrigation channels built at a slight negative slope (-2°) to initiate re-circulation. Plant and animal growth rates as well as water quality were monitored over time to evaluate the conditions which generated the highest economic and environmental benefits. Animal and plant growth were determined in each system as (g $w^{-1}$) and (cm $w^{-1}$) respectively.

Water quality was assessed by measuring total nitrogen TN, NO3, NH4, P04 and orthophosphate every other in both in the aquaculture effluent and influent to elucidate plant and nutrient interactions.

Fish growth rates ranged between the three systems and between experiments did not significantly vary (Multiple regression analysis, $p = 0.43$). Plants irrigated with water from treatments with increasing densities of fish grew consecutively larger and plants grown in larger pots grew to larger sizes than those in smaller pots within the same treatment. Total nitrogen and total phosphorous concentrations remained consistent over time within treatments (Duncans Multiple Range Test, $a = 0.05$) but significantly increased from trts 1 to 2 to 3 (DMR, $a = 0.05$ in relation to increasing densities of fish. Ratios of inorganic to organic nutrient loads for both nitrogenous and phosphorous compounds increased over time and approximately 90% of all nutrients were consistently removed from the effluent as the aquaculture effluent migrated through the potted soil. Approximately 10% water was lost per week as a result of evaporation and plant transpiration (i.e. 90% of the water was re-captured and re-utilized).

Results from this experiment demonstrate that integrative production systems, combining aquaculture with soil based agriculture, is a sustainable and robust means of production. Numerous plant species grew readily when supplied with aquaculture effluent suggesting that RIAA technology can be used to grow both ornamental and food production plant species. The materials necessary to develop RIAA systems are relatively inexpensive, easy to obtain, and easy to maintain. As a result, implementation of this technology in both the United States and lesser developed countries may provide a rapid payback period and high return on investment while minimizing the negative impacts on the environment.

**Green Engineering through Waste Heat Recovery Project Proposal for Johnson Matthey Inc**

Adigun, B., Rowan University

Energy sustainability is very vital for industries and the environment and there has been a global increase in awareness of the benefits. It is broadly applied to fields like Pharmaceuticals, Metallurgy and majorly fields involving process operations, which consists of chemical conversion, production and generation of waste heat.

One of the major by-products of small or large-scale industries is waste heat and a number of equipment has been designed to convert a reasonable amount of this heat for other use through recycling. This project involves design of measures to recover waste heat from a Process Catalyst Technology plant exhaust and directing it for a more useful purpose in the plant. This therefore reduces the carbon-footprint and the use of non-renewable fossil fuel.

**Evaluating Environmental Footprints of Diet Consumption Patterns- Comparing Nitrogen Footprints and Carbon Footprints of Different Foods**

Xue, X., University of pittsburgh
Landis, A. E., Arizona State University

Food supply has become an important contributor to the depletion of natural resources and water quality degradation. Excessive nutrients exported from farming and food processing activities already has resulted in serious hypoxia and eutrophication issues in surrounding ecosystems. Changing food consumption patterns offers a unique opportunity for consumers to lower their personal environmental footprints. Concerned consumers are calling for mitigating environmental burden of food supply. Policymakers and producers therefore require scientifically defensible information about food products and production systems. Recently, significant public concern and scientific attention to environmental impacts of foods mainly focused on their Carbon footprints, few studies have systematically compared the life cycle nutrients flows among different food types. The study of nitrogen and phosphorus inventories for food categories and the potential of reducing nitrification through shifting food consumption patterns have not yet been addressed. Comprehensive comparison of both of carbon footprints and nitrogen footprints over all food categories have not found yet. This study utilizes a life cycle assessment (LCA) approach to quantify the nitrogen flows during food production, processing, packaging and distribution stages over main food types. This study identifies nitrogen footprints of food supply chains among main food categories, compares carbon footprints and nitrogen footprints of different food consumption patterns, and analyzes potential environmental and economic impacts of food consumption shifts. Results show that different food groups exhibit a largely variable nitrogen-intensity. While red meats exhibit the highest nutrient intensity among all food categories, cereals/carbohydrates has the lowest eutrophication potential. Dietary shifting from red meat to cereals/carbohydrates may be an effective way to mitigate eutrophcation impacts resulting from food consumption.

# Kinetics and Modeling of Co-Fermentation Using Saccharomyces Cerevisiae and Pichia Stipitis in Glucose and Xylose Media for Bioethanol Production


## Authors:

Merida-Figueroa, F., University of Puerto Rico at Mayagüez
Saliceti-Piazza, L., University of Puerto Rico at Mayagüez


Ethanol produced from lignocellulosic biomass resources is a fuel with potential to match the convenient features of petroleum, but reducing substantially the emissions of greenhouse gases in comparison with fossil fuels. Using either pure ethanol or in mixed blends with gasoline, the fuel octane is boosted, tailpipe emissions of CO are reduced, and fuel combustion is improved. A wide variety of lignocellulosic biomass is available in our planet, and its major components, cellulose and hemicellulose can be used to obtain 6- and 5-carbon carbohydrates. To achieve this, biomass must be pre-treated and hydrolyzed. The resulting carbohydrates can be converted to ethanol following several biochemical pathways in which microorganisms play an important role as biocatalysts. This work features the fermentation of mixtures of glucose and xylose as the main components of cellulose and hemicellulose, respectively; to obtain ethanol using two different native yeast strains ofSaccharomyces cerevisiae and Pichia stipitis, whose efficiency to ferment glucose and xylose has been demonstrated in previous studies. For P. stipitis, efficiency was also studied in our preliminary screening of xylose-fermenting yeast strains including Pachysolen tannophilus and Candida shehatae . We are focused in determining a series of experimental kinetic parameters, including specific cell mass growth rate, substrate uptake, production formation and yield coefficients to evaluate ethanol production, to mathematically model sugar consumption and ethanol production. Modeling of the fermentation processes through adequate mass balances and biokinetics to match with experimental results for both individual and mixtures fermentations will become a very important tool for the design of efficient biomass fermentation processes.

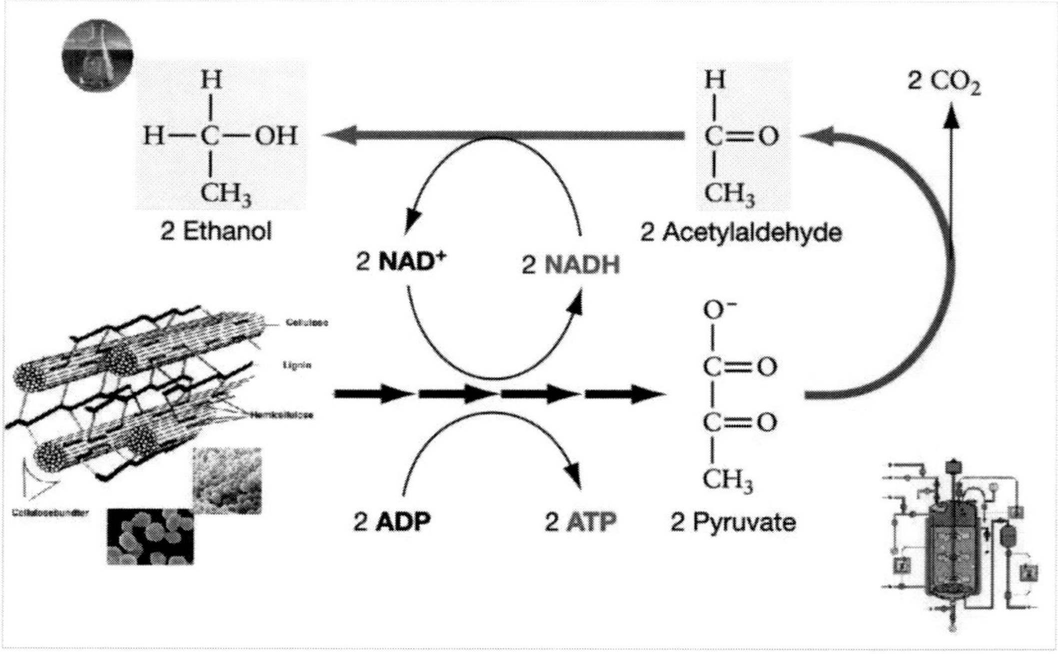

**Supercritical CO2 Hydrolysis and Explosion as Pretreatment of Guayule Bagasse for Fermentation Feedstock**

Srinivasan, N., University of Akron
Ju, L., University of Akron

Guayule is a desert shrub grown in the dry regions of Arizona and California for commercial production of hypoallergenic latex. Its resins also have unique applications. These products are non-allergenic to most people unlike the products made from the rubber from b. Latex and resins, however, make up only about 20% of biomass. Converting the waste bagasse to biorefinery feedstock for value-added products makes good economic sense. A supercritical CO2-based process had been developed for extracting resins and rubber from guayule. In this study, we demonstrated the feasibility of using a supercritical explosion-hydrolysis procedure, as part of the integrated process, to prepare/pretreat the bagasse for subsequent enzymatic and microbial conversion. Our approach used supercritical CO2 and moisture for mild hydrolysis and subsequent explosion, created by sudden pressure release, to loosen and expose the lignocellulosic structure for enzymatic hydrolysis. The pretreatment involved: adding water to the bagasse, raising system temperature, pressurizing using supercritical CO2, holding the system for a period of time, and exploding the bagasse. The pretreated biomass was subjected to enzyme hydrolysis. The yields of released sugars were used as the indicators for pretreatment effectiveness. Comparison studies showed that this method was more effective than other techniques such as acid hydrolysis and delignification. The enzymatic hydrolysates were also tested on Trichoderma reesei Rut C-30. No inhibitory/toxic effects were apparent in terms of cell growth, sugar consumption, and cellulase and xylanase production. Optimizing the operating parameters like temperature, pressure, moisture and duration of hydrolysis can increase the sugar yield. A factorial design was performed to find the optimum conditions of temperature, pressure, moisture and duration of hydrolysis for maximum sugar yields from the supercritical CO2 pretreatment of the guayule. The yields of sugars released from the enzyme hydrolysis were used as the response variable in the central composite design used. The optimum glucose yield was 55% and pentose yield was 60% from enzyme hydrolysis. XRD analysis of the bagasse samples pretreated under different supercritical conditions confirmed the relationship between the crystallinity index and the ease of hydrolysis. SEM pictures showed the effect of explosion and hydrolysis. SEM and XRD data correlations provided additional insight into the pretreatment process.

**Concentrated Solar Cooking and Heating System**

Valmiki, M., University of Arizona

The second law of thermodynamics tells that energy, in many different modes, has quality, which measures the usefulness of the energy. It is easy to understand that electrical power is the energy of the highest quality, since electricity is often generated by concentrating larger supplies. Using electricity for heating is a poor and inefficient use of dense energy since it is used in an inefficient way, generally dissipating largely unused to the environment. Using solar energy for heating, however, is a highly economic use of energy. Solar radiation travels to the Earth as heat and may be directly used for heating purposes in extremely efficient energy life-cycles.

Although solar heating has long been recognized as a cheap, effective method, there are not many well-studied and designed products available to the public. One typical example is the traditional solar cooker or solar oven. To cook in front of a reflective solar collector, people are often subject to strong and concentrated sunlight, causing considerable safety issues involving both burns and blinding. Furthermore, traditional solar cookers do not allow for much control of surface temperatures or sun location tracking. Finally, parabolic and box solar cookers can only be used outside due to their inability to store and transfer the heat indoors.

A student team was assembled according to the system developed by the instructional staff for the University of Arizona's engineering senior design course. The team consisted of one optical and five undergraduate mechanical engineering students. In addition to the design and development tasks required for the P3 timeline, the course demanded several presentations, reports, design processes, and other minor deliverables. The fall semester was dedicated to design considerations while spring was reserved for prototyping and testing.

It is expected that a safe and versatile solar cooking and heating device which remedies these issues will garner significant interest worldwide. A user-friendly, safe, controllable system could have many potential applications. The opportunities of marketing such a sustainable technology are enormous both inside and outside of the US. Phase I of this project was to develop a method of cooking and heating using only the sun, in order to improve upon common solar cooking technology.

The research is in early stages and only one prototype has been completed. The prototype is a proof of concept for both outdoor and indoor applications.

Figure 1 - Thermal loop schematic

The system principles are relatively simple to understand. Incident sunlight is concentrated by a large Fresnel lens by a factor of about 750. This focal point is small and stationary throughout the day. A fixed focal point allows for a small collection area to maximize efficiency. This is achieved by dual-axis, manual sun tracking about the focal point.

The beam of light is of very high intensity and can be dangerous to users. However, the system can be made very safe by isolating the concentrated light with a curtain or shield.

The focal point heats a small, blackened collection surface adjacent to the outdoor cooking surface. The surface is part of a chamber filled with mineral oil, the working fluid. For outdoor cooking purposes, no mass flow is enforced, keeping the heat in the chamber. The oil circulates passively, transferring the heat to the cooking surface evenly. Alternatively, for indoor use, the oil is pumped through a piping loop using an internal gear pump. The oil is transported 6 feet to an indoor heat exchanger embedded in a secondary cooking surface, representing the potential of indoor application. The pump is powered by a DC motor, but could be easily powered with photovoltaic conversion.

Figure 2 - EPA P3 conference and testing

Due to time constraints of the students, only several tests were conducted, but the results are promising. The outdoor surface reached 290°C, as hot as a kitchen stove. At this point the oil reached its boiling point and a load had to be added so that it did not boil and break down. It is expected that the outdoor surface temperature could go higher if oil were not present in the chamber, restricting the maximum temperature.

At 140 minutes flow through the loop was initiated to test the indoor performance. With throttling of the flow rate the indoor surface temperature reached 150°C.

The team was able to pan-fry potatoes and boil broccoli on the outdoor surface. The indoor surface got hot enough to cook hot dogs to a safe temperature (170°F) directly on the metal. Insulating covers to enclose both the cookware and cooking surface could help trap heat. A simple glass cover could suffice. These could be used to create a baking environment.

Figure 3 - Full performance curve

The first full testing yielded results that could be used to calculate both the indoor and outdoor efficiencies. Unfortunately, testing was performed on a windy and partly cloudy day. As such, it is difficult to determine the solar radiation flux without proper equipment. Also, the free convection was enhanced by the wind. It was estimated that the solar radiation flux on this day was 600 W.

Using heat transfer methods, the power output at the indoor surface was calculated to be 106 W. This results in an estimated efficiency of 18%. The power output from the outdoor surface testing was calculated as 465-500 W. This results in a range of estimated efficiencies of 77-83%. This efficiency range matches intuition. For indoor cooking to be viable, the indoor efficiency must reach this higher range.

Time will be needed for development into a marketable, fully-implementable product.

Already, several organizations, both commercial and academic, have expressed interest in collaboration and a provisional patent has been filed.

The ideas presented here could be applied in many ways. Especially intriguing are commercial cooking operations, outdoor models, and small, residential scale solar thermal energy applications. Modular systems of multiple lenses could provide heat for a large cooking operation by heating large griddles or cooking oil. Outdoor models could be applied in developing countries, roadside vendor operations, and backyard cooking. If cost effective, such stoves could reduce the need for cutting of trees for fuel in rural regions where electricity and gas are not available.

Residential scale solar thermal systems could provide heat for cooking, air conditioning, or water heating. In concert with a small thermal energy storage unit, the system could provide usable heat for nighttime use. Along with automatic tracking, the unit could be installed on rooftops for full sun exposure and aesthetics. Piping could be installed within walls by professional contractors just as water pipes are. The system could run all day while the owners are at work, providing heat when they need it.

An economic study has not been done, but it is projected that the cost would be about $400 for an outdoor unit and about $2000 for an indoor unit. The cost of an indoor unit could be included in home construction.

The prototype is not ideal, as is the case of most first-generation models. Shortcomings include excessive weight and indoor surface temperature outputs. Redesign processes should take some considerations into account.

Different outdoor and indoor surface designs could be numerically determined, fabricated, and tested. In this way, volumes, channel design, collection surface geometry, and materials could be optimized. One issue that needs work is the transfer of heat from the collection surface to the oil. No computational modeling of the heating chamber was performed, so there is much room for improvement. By optimizing the chamber geometries, the oil flowing through can be heated more efficiently. Fins within the chamber could be part of this solution.

The efficiency loss due to the length of piping could also be minimized. As the oil flows through the piping to and from the secondary surface, heat is lost. Insulation choices were not optimal and much of the insulation is just taped on. Perhaps using the same type of insulation used in solar thermal power plants could be beneficial.

Perhaps, if more energy supply is needed, a larger lens or multiple lenses could be implemented. An aperture control of the lens could help regulate the energy input for enhanced temperature control. A control system could be implemented to regulate the temperatures based on desired user temperature inputs.

Continuous, automatic control using small, solar powered motors can enhance the system by eliminating the need for manual tracking. A control system using a light sensor or sunangle database could accomplish this.

The global demand for energy has led the world to a momentous time in history, yet an opportunity of unprecedented magnitude exists. New, sustainable energy technologies must be integrated into the current system, and this project could be one of those many important possibilities. Technologies such as this concentrated solar cooking system are needed for the continued prosperity and health of the planet and its peoples.

**Forging New Links:Toward Sustainable Supply Chain Management**

Fiksel, J., The Ohio State University

Originally presented on: 8/11/2009 8:30:00 - 9:10:00

Corporate sustainability requires awareness of the full product life cycle, ranging from the conduct of upstream suppliers to the disposition of obsolete products. Leading companies have integrated environmental and social responsibility expertise into cross-functional teams throughout the supply chain, thus enhancing profitability, resource productivity, innovation and growth. For example,?environmentally preferable purchasing? has influenced the development of benign products that are energy efficient and avoid the use of harmful substances. Likewise, anticipating safety and security risks and developing contingency plans has enhanced business continuity and resilience.

The speaker will provide an overview of best practices in sustainable supply chain management, based on a study funded by the Global Environmental Management Initiative, a prominent industry consortium. The business processes addressed include supplier relationship management, materials management, order fulfillment, product development, and returns management. Companies that participated actively in these studies include 3M, Anheuser-Busch, Dell, Dow Chemical, Duke Energy, DuPont, Eli Lilly, FedEx Express, HP, Intel, Kodak, Motorola, Novartis, Pfizer, Procter & Gamble, and Texas Instruments. Their experiences illustrate how corporations can simultaneously meet stakeholder expectations with regard to sustainability and superior business performance.

Finally, the speaker will present guidelines for disruptive environmental innovation, which has the potential to achieve quantum improvements in the ecological footprint of global supply chains. Based on a new book, "Design for Environment: A Guide to Sustainable Product Development," the guidelines encompass dematerialization, detoxification, revalorization, and capital preservation. Such innovation is the only plausible pathway to meet the challenges of climate and natural resource protection while realizing economic prosperity and social equity.

**A Billion Tons of Biomass: Toward a Sustainable Biomass Feedstock Infrastructure**

Richard, T., Pennsylvania State University

Originally presented on: 8/11/2009 9:10:00 - 9:35:00

The US is embarked on a trajectory to produce up to 30 percent of our transportation fuels from biomass by 2030. As one indicator of the scale of this challenge, it is useful to benchmark the USDA-DOE estimate of the biomass required, one billion tons, against our current agricultural sector. Current U.S. production of grains, fruits, vegetables, hay, and even pasture grasses combine to only 800 million tons. To achieve this scale we will need the equivalent of a second agricultural infrastructure, bigger than that the current one, arising from nowhere in the next twenty years. Imagine all the workers, equipment, and other infrastructure investments that will require. In the U.S. we have already seen railcar shortages in the last few years due to rapid increases in corn ethanol production. With even more rapid scale-up of cellulosic technologies predicted, we can expect substantial shortages of railcars and trucks, as well as greatly increased traffic on the rural roads and short-line rails that will carry this feedstock about. For new, high yielding bioenergy crops we will need new harvest equipment, and to contain transportation costs we will need new methods of densification before transport. To insure a large biorefinery has a secure a year-round feedstock supply, we will need massive storage facilities, perhaps satellite pre-processing centers, and a tightly choreographed strategy for moving all that material around. And beyond these technical considerations are the human factors ? enthusiastic producers, cooperative neighbors, and perhaps 200 new workers in the feedstock supply chain for every 50 keeping a biorefinery hum. To understand and prepare for these challenges, we need to do our homework now ? inventorying feedstock, talking to landowners, developing supply-price curves, and investing in the producer co-ops, equipment companies, transportation infrastructure and the people that this industry needs. If we desire a future where a substantial fraction of our energy comes from biomass, it is increasingly clear that we need immediate and substantial investments in feedstock infrastructure to enable that future growth.

# Sustainable Supply Chains: A Framework for Implementation

## Authors:

Goldsby, T., University of Kentucky
Badurdeen, F., University of Kentucky
Metta, H., University of Kentucky
Wijekoon, K., University of Kentucky
Stoval, C., University of Kentucky
Jawahir, I., University of Kentucky
Iyengar, D., University of Kentucky

Abstract: Sustainable supply chain management requires an integrated approach that focuses on the total product life-cycle from pre-manufacturing, manufacturing and use through post-use stages from a triple bottom line perspective; economic benefits, environmental protection and societal well-being. Achieving sustainability in the supply chain therefore requires adopting a closed-loop approach to material flow. The 6R methodology that extends the previous 3R's (reduce, reuse, recycle) to include the additional steps of recover, redesign and remanufacture provides an ideal platform for approaching sustainable supply chain design and management. However, none of the existing supply chain management frameworks embrace sustainability from a holistic perspective nor do they provide the metrics needed for the evaluation of supply chain sustainability. This paper will present an integrated approach to sustainable supply chain management based on the 6R methodology, provide a comparison of existing supply chain frameworks and their suitability from a sustainability point of view and elaborate on the road map that must be followed to design and manage sustainable supply chains.

**Supply Chain: The Critical Enabler for Meeting Corporate Sustainability Objectives**

Callieri, C., AT Kearney

Originally presented on: 8/11/2009 10:00:00 - 10:25:00

Chemical Industry executives are recognizing the importance of implementing sustainability and social responsibility (SSR) initiatives for their organizations. However, as with any emerging topic there is some debate or confusion about the definition and scope of sustainability or corporate social responsibility. In this brief presentation we will address why these two areas are intimately connected and why chemical industry companies should start with a broad definition of sustainability including economic, environmental and social aspects, in order to define its own corporate definition of sustainability. In a globalized world where supply chains have become increasingly complex and important to production, the Supply Chain will become the litmus test of sustainability for any company. We will outline a guide to evaluating the sustainability of your company supply chain to identify gaps and opportunities, as well as enable the development of an appropriate Supply Chain Sustainability Strategy that is aligned with the broader corporate strategy.

**A Sustainable Bio-Supply Chain Begins with Sustainable Agriculture Cropping Systems**

Karst, M., Entira

Originally presented on: 8/11/2009 10:45:00 - 11:10:00

This presentation will illustrate how industrial processors can partner with the agricultural community to develop a sustainable bio-supply chain using the pillars of environmental health, economic profitability and stakeholder equity.

Key points will include:

? the implications of an alternative crop for existing cropping systems

? the agronomic and environmental considerations of an alternative crop

? the economic risks for growers including capital needs, loss of open market flexibility, loss of risk management tools, and the overall uncertainty of a new crop

? the range of gross revenues that the alternative crop must return to be competitive with the opportunities with traditional crops

? the storage and logistics issues that must be addressed to make an alternative crop viable

? How might industrial processors partner with growers to mitigate these risks for both parties and make alternative crops viable feedstocks for a bio-supply chain?

**Preshipment Package Testing Validate Sustainability**
Church, E., ISTA
Originally presented on: 8/11/2009 11:10:00 - 11:35:00

During transit all packages are submitted to the environmental hazards of shock, vibration, dynamic compression and atmospheric elements. As package designers work on making packages more sustainable they can not forget about these forces. Environment gain maybe completely lost if damage occurs to the product and sometimes just the package. The idea is to engineer a package that protects the product, but does not over-package it.

Learn how to validate sustainable packaging efforts by developing preshipment package tests to simulate the distribution environment. Hear case studies of success stories.

**Sustainability Approach Delivers End to End Supply Chain Solutions**

Johnson, W., Procter & Gamble

Originally presented on: 8/11/2009 11:35:00 - 12:00:00

P&G recently announced significantly increased targets for its 2012 sustainability goals. The updates reflect the company's continued commitment and progress in improving the environmental profile of P&G's products and operations and bettering lives of children in need through its Live, Learn and Thrive? corporate social responsibility programs.

Specifically for operations, the goal was increased to 20 percent reduction (per unit of production) in carbon dioxide emissions, energy consumption, water usage and disposed waste from operations, leading to a total reduction over the decade of at least 50 percent. These results are from economically smart and sustainable designs throughout our supply chain.

# Global Climate Change and the Technology Challenge

## Authors:

Princiotta, F., USEPA

Anthropogenic emissions of greenhouse gases, especially carbon dioxide, ($CO_2$), have led to increasing atmospheric concentrations,
very likely the primary cause of the 0.8 °C warming the earth has experienced since the industrial revolution.

With industrial activity and population expected to increase for the rest of the century, large increases in greenhouse gas

emissions are projected, with substantial global additional warming predicted. Of particular importance are rapidly developing

countries such as China and India, whose emissions are growing rapidly in the power generation, transportation and

industrial sectors as their economies grow and standards of living improve.

The paper discusses forces driving ($CO_2$) emissions, the levels of emission mitigation needed to constrain warming to a

given value, a concise sector-by-sector summary of mitigation options, and research and development (R&D) priorities. In order to constrain global warming in 2100 to below about 2.5+/-0.7 °C, the current annual 3% ($CO_2$) global emission growth

rate must be transformed to a 1 to 3% annual declining rate for decades, as soon as possible. This will require a rapid and

radical transformation of the world's energy system. The current generation of energy technologies is not capable of achieving

the mitigation required. Recent publications were used to relate the implications of a 3.7 trillion-ton mitigation program

needed to constrain warming to the key energy sectors and the technologies within those sectors that can make major mitigation

contributions. It is concluded that an aggressive, cost effective mitigation program relying on existing technologies is

capable of mitigating only between about 25% and 45% of the required ($CO_2$). Therefore, in the absence of fundamental lifestyle

changes, new technologies are required for the key energy-related sectors: power generation, transportation, industrial

production, and buildings. The power-generation sector and transportation sectors are particularly important, since they are

projected to grow at relatively high rates, driven especially by China and other actively developing countries.

It will be necessary to substantially upgrade and accelerate the current worldwide RDD&D effort. Given the monumental

challenge and uncertainties associated with a major mitigation program, it may be prudent to consider all available and

emerging technologies. This suggests that fundamental research on energy technologies in addition to those currently in

advanced stage of development, be part of the global research portfolio, since breakthroughs on today's leading-edge technologies

could yield tomorrow's alternatives.

It will be necessary to substantially upgrade and accelerate the current worldwide RDD&D

effort. Given the monumental

challenge and uncertainties associated with a major mitigation program, it may be prudent to consider all available and

emerging technologies. This suggests that fundamental research on energy technologies in addition to those currently in

advanced stage of development, be part of the global research portfolio, since breakthroughs on today's leading-edge technologies

could yield tomorrow's alternatives.

In light of this monumental mitigation challenge, geoengineering options, i.e., those that aim at fundamentally changing

the Earth's heat transfer characteristics, should be evaluated and, if viable, seriously considered for their potential to cool the

planet and allow more time for the necessary energy transformation.

Finally, availability of key technologies will be necessary but not sufficient to limit ($CO_2$) emissions. Since many of

these technologies have higher costs and/or greater operational uncertainties than currently available carbon intensive technologies,

robust regulatory/incentive programs will be necessary to encourage their utilization

**Carbon Negative Biomass Chemical Looping (BCL) Process for Hydrogen and Power Generation**

Fan, L. S., The Ohio State University
Li, F., The Ohio State University
Zeng, L., The Ohio State University
Kim, H. R., The Ohio State University
Sridhar, D., The Ohio State University
Wang, F., The Ohio State University
Tong, A., The Ohio State University
Originally presented on: 8/11/2009 9:10:00 - 9:35:00

A novel, carbon negative Biomass Chemical Looping (BCL) process is discussed. Utilizing iron oxide as the oxygen carrier, the BCL process produces hydrogen and electricity from cellulosic biomass, lignin-rich non-fermentable residues, and agricultural residues with 100% $CO_2$ capture. From life cycle analysis (LCA) standpoint, the process is carbon negative with a net $CO_2$ emission of less than -1.3 tons per ton of biomass converted. Preliminary experimental results indicate that the BCL process concept is feasible. A comprehensive ASPEN Plus® model taking into account both the preliminary experimental data and thermodynamic restrictions is developed to evaluate the performance of the BCL process. Based on the ASPEN Plus® simulations, close to 70% process efficiency (HHV basis) can potentially be achieved for hydrogen and electricity co-production from cellulosic biomass. The high energy conversion efficiency of the BCL process results from the process intensification coupled with energy management optimization.

**Solar Cell Technologies for Non-Fossil Power Solutions**

Singh, V., University of Kentucky

Originally presented on: 8/11/2009 9:35:00 - 10:00:00

Several solar cell technologies including single crystal silicon, polycrystalline silicon, thin film cadmium telluride and copper indium diselenide have reached the commercial production phase. Other technologies like dye-sensitized solar cells, nanocrystalline devices and organic semiconductor solar cells hold the promise of flexible, lightweight designs and adaptability to large scale production at low cost; these are in the research and development stage. All have a potential role in meeting the non- fossil power needs of human kind. Device designs, current performance levels, technical challenges and future potential of these technologies will be described and compared against each other.

Bio-summary (Vijay P. Singh)

Dr. Vijay P. Singh (M'99) received the B. Tech. degree in electrical engineering from I.I.T. Delhi in 1968 and M.S. and Ph.D. degrees in electrical engineering from the University of Minnesota in 1970 and 1974 respectively.

He began his research career in solar cells at the Institute of energy Conversion, University of Delaware in 1974. He joined the microelectronics industry in 1976 first as a Research Engineer and later as Section Head and Manager of Materials and Device Research, for Photon Power Inc., El Paso, Texas. In 1983, he joined the faculty of the Department of Electrical and Computer Engineering at the University of Texas at El Paso as an Associate Professor. He was promoted to the rank of Professor in 1990 and in1993 was named Schellenger Chair and Director of the Electronic Devices Laboratory. He also served as Associate Dean for Research for the College of Engineering till 2000. In 2000, he joined University of Kentucky in Lexington where he is a Professor, Earl Parker Robinson Chair in Engineering and the Director of the Center for Nano-Scale Science and Engineering. He is the author of more than 100 journal and technical articles, patents and book chapters. His research interests are in the areas of solar cells, nanoelectronics, microelectronics, electronic devices and materials, thin-film technology and electroluminescent displays. Dr. Singh is active in several professional and technical societies including IEEE and APS and has served on the organizing committees and chaired technical sessions for a number of international conferences related to nanoelectronics, solar cells and display devices.

**Engineering Sustainabilitiy Development and Its Application in Fuel Cell Systems**

Mannan, M. S., Mary Kay O'Connor Process Safety Center
Ashfaque, S. A., Mary Kay O'Connor Process Safety Center
Guo, Y., Mary Kay O'Connor Process Safety Center
Originally presented on: 8/11/2009 10:00:00 - 10:25:00

Fuel cell system is considered as one of the most important sources of clean energy in the near future to meet the challenge of environment friendly global energy demand, however its sustainability as a fuel has not been completely established, including environmental, economical and safety impacts. In this paper, a sustainability analysis is performed for Fuel Cell systems compared with other current and future potential energy sources. Most of the sustainability studies have been concerned so far only with the environmental impact and less rigorous methods have been used in other studies to quantify the sustainability of energy systems. The methodology used in this study ensures a Sustainable Development of a process not only with respect to environmental impact, resource consumption but also with respect to societal and economic impacts. It incorporated indicator-based analysis, which allow provisions for benchmarking for alternative processes and simplifies priorities for several sustainability elements of the processes. Indicators were selected from mostly seen parameters in manufacturing process systems. To integrate those indicators with the Fuel Cell energy systems the formulations were modified to adopt it with the energy systems. Three of the major prospective fuel cell systems PAFC, SOFC and MCFC were considered in this study to compare their sustainability indicators. Some of the difficulties can be encountered to quantify the performance of the systems are due to lack of sufficient process specific data, scarcity of necessary information and satisfy the needs for decision makers. To overcome these difficulties a decision making tool was used to compare and set priorities to these indices of different fuel cell technologies. Since the performance indicators analyzed for this system varied widely, a Multi Criteria Decision Making (MCDM) process was used in this purpose. Analytical Hierarchy Process (AHP) is chosen as a MCMD tool which was applied to compare the indicators of different fuel cell process alternatives under study. The Sustainability indicator priority score and each individual alternative's performance score together are used to determine the most sustainable alternative.

**Using MARKAL Model to Evaluate Factors Influencing Low Carbon Power Generation**

Loughlin, D., EPA

Originally presented on: 8/11/2009 10:45:00 - 11:10:00

There is considerable uncertainty in how U.S. power generation will evolve over the next decades. Factors anticipated to influence this evolution include the availability and cost of fossil fuel resources, the capital and operating costs of competing electric generation technologies, the level of public acceptance of nuclear power expansion, the viability and timing of large-scale carbon capture and sequestration, the ability to rapidly increase renewable capacity, the effect of research and development programs on improving technology cost-effectiveness and availability, the impact of current and future environmental regulations, and the role of climate policy on putting a price or limit on carbon dioxide emissions.

The U.S. EPA's Office of Research and Development (ORD) is using the MARKAL energy system model to evaluate how various assumptions about these factors may affect the evolution of the U.S. energy system. MARKAL represents the principle components of the energy system, including current and projected energy supplies, energy demands, and the technologies that produce and consume energy. The model uses this information to identify cost-effective technological pathways for meeting energy demands through 2050. The relatively quick runtime of MARKAL (5 minutes for the national model and 35 minutes for the regionalized model) facilitates its use in sensitivity and uncertainty analyses. In this presentation, a nested parametric sensitivity analysis application is described, providing insight into the relative impact of various assumptions on the evolution of low carbon power generation.

**Coal and Biomass to Electric Power and Fuels**

Katzer, J., MIT

Originally presented on: 8/11/2009 11:10:00 - 11:35:00

Coal is a growing component in power generation and is on an upturn

globally because it is cheap and abundant. Coal generates 50% of U.S.

electricity but is facing increasingly stiff opposition because of its image of being ?dirty? and because of public/political concern over CO 2

emissions. CO 2 can be captured from PC flue-gas for geologic storage,

but at a cost. Integrated Gasification Combined Cycle (IGCC)

generation can capture CO 2 at lower cost and can be very clean with

respect to criteria emissions. Oxy-fuel combustion has potential for

power generation with CO 2 capture but requires commercial development.

At the same time, there is increasing interest in producing liquid transportation fuels from coal, but CO 2 emissions are twice those of

petroleum-based fuels, unless geologic CO 2 storage is used. Using

biomass and biomass/coal mixtures can reduce life-cycle CO 2 emissions

and provide a path into sustainability. The performance and costs of

these technologies are compared on a consistent basis, without CO 2

storage and with CO 2 storage. With biomass, CO 2 emissions can be

markedly reduced and can be negative, but power and fuel costs are

higher. Base-load electricity generation from coal with CO 2 capture

and storage is economically competitive with power from wind and new

nuclear power plants and is very clean. Integrating power generation

and liquid fuels production can provide enhanced opportunities to economically meet power and fuel demands and decarbonize significant portions of these two sectors. However, to accomplish this requires

successful commercial demonstration of geologic storage. At the same

time, there do not appear to be any irresolvable scientific or technical issues, or economic showstoppers to long-term geologic storage of captured $CO_2$ with safety and efficacy.

**Advanced Carbon Management Technology and Development Including Carbon Capture and Storage**

Cugini, A., U.S. Department of Energy, National Energy Technology Laboratory
Syamlal, M., National Energy Technology Laboratory, U.S. Department of Energy
Originally presented on: 8/11/2009 11:35:00 - 12:00:00

Effective carbon management technologies can greatly improve the efficiency of energy conversion from fossil fuels, while at the same time significantly reduce the impact on the environment. This paper discusses efforts to develop these advanced carbon management technologies, including improving the efficiency of current and developing conversion processes, carbon capture and storage, carbon dioxide utilization and reuse, and reducing the carbon intensity of fossil energy use. Technologies for all areas will be discussed, with a focus on current efforts related to carbon dioxide capture and storage. A review of techniques to efficiently capture carbon dioxide from pulverized coal and gasification power plants while attempting to maintain the efficiency of these power plants is presented. Advanced power cycles, designed to effectively separate CO2 into a sequestrable stream, are discussed. Geological storage options for CO2, including depleted oil and gas reservoirs, unmineable coal seams, and deep saline formations, are discussed. These storage options will be described along with emerging concepts to "recycle" CO2 into useful chemicals or fuels, using renewable energy to eliminate carbon emissions.

# Short and Long Term Sustainability of Technologies for Liquid Biofuels

## Authors:

Wenzel, H., University of Southern Denmark

Two main roads to biomass conversion to transport fuels exist; the thermochemical and the biochemical conversion. Several fuels for road transport are in full scale production, mainly from the biochemical route in the form of first generation bio-ethanol and biodiesel. A second generation biodiesel production based on rendered animal fat was established in full scale in Denmark this year, and so was a large scale/semi full scale second generation bio-ethanol from straw.

The thermochemical conversion route has promising technological perspectives, and the prospect is that transport fuels produced from ligno-cellulosic biomass via the thermochemical route will be available on the market within a 5 year period. The key technologies of gasification and liquefaction are well known using other precursors than biomass. The technological perspectives are one thing, however, the feasibility and sustainability issues another. To assess the sustainability of biofuel technologies, one has to consider some proportions.

Proportions in demand and supply of biomass

Earth land coverage is around 15 Gha of which 5 Gha is in use for agricultural production either as cultivated land for crop production (around 1.5 Gha) or pastures for animal production (around 3.5 Gha). The remaining 10 Gha are either biophysically unsuited for cultivation or cultivable, but standing as uncultivated nature, i.e. primary forest, wetlands, and other biome categories.Till now, the cultivated part of the land has largely been cultivated for food and animal feed production. But many new customers to biomass resources enter the scene:

electricity, heat, road transportation, aviation, polymers, organic bulk chemicals. A quick look at the proportions and the magnitude of these newcomers, compared to our agricultural sector as we know it, illustrates how big they are: With the average daily diet of around 2700 kcal per person, the total calorific food intake is around 25 EJ per year by the world's population. The global fossil fuel consumption today is around 400 EJ per year, i.e. 16 times larger than the energy content of the world's food intake.

The energy content of the crops deriving from agricultural land is, of course, bigger than the 25 EJ/year, due to the losses in the supply chain for food, and based on data from FAO Statistics Division in 2007, the gross energy production in agricultural crops today, thus, amounts to an estimated 100 - 150 EJ/year. Acknowledging that also biofuel/bioenergy production imply losses in the supply chain, the proportion is, thus, that using biomass for all energy demands, today satisfied by fossil fuels, would require four to five times the present energy output, on top of the existing agricultural production.

Moreover, consumption increases in both energy and food tend to exceed yield increases, for the food part not least due to a growing demand for meat. With present trends, therefore, yield increases do not seem to close the gap, on the contrary. Looking at possibilities for extension of agricultural land, leading geographers have found that earth can provide a maximum of 120% new cultivable land, most of which is found in tropical South America and Africa. The figure is theoretical, and actually cultivating this land would imply deforestation including violation of nature preservation, and the realistic magnitude of new land cultivation is much lower. A recent study commissioned by the Danish Ministry of Food and Agriculture finds the sustainable potential to be in the order of 30-40% new cultivable land only. Moreover, forest and other nature types have in many cases sequestered much more carbon than the agricultural land following the cultivation, and the release of carbon due to cultivation may be very high compared to the subsequent carbon offsetting by the crops substituting fossil fuels, and ?carbon pay-back periods' as high as 400 years are reported. The conclusion of this rough look at proportions is that while we need five times more cropland in order to fully replace fossil fuels by biomass, we can at maximum double our cropland, and we can do far from that without jeopardising nature conservation laws and without carbon releases exceeding carbon savings.

This quick scan of proportions matches well the fact that most scientific studies of biomass potentials estimate the maximum potential to be in the area of 20% of energy consumption around 2030, some of the most optimistic around 50%, and the ones looking not only on maximum potential but also at economic feasibility around 10%. Looking at the biomass residue potential only, being the biomass that can be extracted without competition with food production, the total population of studies estimate a potential in the interval of 15 ? 96 EJ/year, or around 2 ? 13% of projected fossil energy consumption in 2030.

The short term perspective

These proportions in mind, the issue of competition for the biomass and the need for prioritizing it become evident. For several decades ahead, we still depend heavily on fossil fuels, and we can only replace these to the extent and at the speed that alternatives become available. Any large-scale prioritisation of biomass for one purpose will imply a loss of alternative uses of the same biomass. The use of biomass for transport fuels shall, thus, be seen in the light of competition for and prioritisation of the constrained land and biomass resources, not only in relation to food and feed demands for biomass and land, but in relation to all new potential uses of biomass and arable land.

The studies using this proper holistic comparative perspective unambiguously show that more is lost than gained when prioritising biomass for transport biofuels at the expense of heat and power. Looking at both $CO_2$ reduction and fossil fuel savings, the conversion of biomass to heat and power or its use in chemistry can imply from fifty up to several hundred percent higher savings per ton of biomass and/or per hectare than biofuels, depending on the type of biofuel.

In order to compete, the success criteria for any use of biomass for energy are: high crop yield, high conversion efficiency, and high cost efficiency. Acknowledging this, it is easy to explain the bad environmental performance of the liquid biofuels: Bio-diesel from vegetable oil implies a very low crop yield (temperate regions), up to 3 times lower yield of dry matter per hectare compared to other energy crops. Bioethanol, both 1st and 2nd generation, has a low energy-conversion efficiency being around 50-70% including the by-products. The reason is that bioethanol suffers from several significant conversion losses: pre-treatment (2nd generation), low metabolic conversion to ethanol, subsequent need to dry residual unconverted matter, and distillation to separate ethanol from water. The same will hold true for biobutanol, and also the thermochemial conversion routes imply significant losses. When using biomass to substitute fossil fuels in heat and power production on the contrary, a substitution efficiency close to 100% can be achieved.

On the short term, therefore, when biomass can be exchanged for oil or natural gas in the heat and power sectors, this is by all judgements a better option, then using the saved oil and gas for transport purposes. Oil and gas can in this way be saved for road transport and oil can be saved for aviation by further converting road transport to natural gas. Assuming properly designed incentives for the use of biomass for energy, conventional market forces would pick this strategy as the winner ? like would the pure environmental concerns.

The long term perspective

When oil and gas are no longer used in stationary applications, and biomass thus no longer can be exchanged to oil or gas in the heat or power sectors, alternative fuels for road transport are justified. At that time, the electric car for road transport is believed to be available being superior in terms of both energy efficiency and environmental impacts. When oil is no longer used in stationary appliances or in road transport, and oil, thus, no longer can be saved for aviation there, alternative aviation fuels are justified. At that time, BTL jet-fuels will have to compete with GTL and CTL as well as other routes to fuels, e.g. wind and solar based electrohydrolysis and methanol production from H2 and CO2. In case BTL proves competitive, aviation is judged to be a priority customer for the available biomass. A trend based projection of aviation fuel consumption to 2030 estimates an aviation fuel consumption of around 25 EJ/year equivalent to 40-50 EJ/year of biomass, if bio-based. Projected chemical feedstock is likewise estimated around 50-60 EJ/year in 2030. These two sectors together can, thus, take more than the estimated maximum non-food biomass potential in 2030, and they are believed to have higher priority for carbon based fuel and feedstock than road transport. So has a percentage of biomass as a storable fuel to buffer electricity systems.

It is, thus, difficult to see the justification of the liquid biofuel technologies, as we know them, on the short term as well as the long term.

**Fossil Fuels, Biofuels, and Electricity -- Well-to-Wheels Energy Use and Greenhouse Gas Emissions Analyses**

Wang, M., Argonne National Laboratory
Originally presented on: 8/11/2009 14:10:00 - 14:35:00

To reduce national dependence on imported oil and reduce greenhouse gas (GHG) emissions, research on alternative transportation fuels and vehicle technologies has gained increased support from governments and private organizations. In order to thoroughly compare the impact of alternative fuels and advanced vehicle technologies on energy and environment, fuels and vehicles need to be examined throughout their life cycles. Argonne National Laboratory (ANL) developed the Greenhouse gases, Regulated Emissions, and Energy use in Transportation (GREET) model to evaluate the full fuel cycle or well-to-wheels (WTW) energy use and GHG emissions of alternative fuels/vehicle technologies. The vehicle technologies considered in this analysis include internal combustion engine (ICE), regular hybrid electric vehicle (HEV), plug-in hybrid electric vehicle (PHEV), and fuel cell vehicle (FCV). The investigated fuels include gasoline and diesel from conventional oil, ethanol from corn and cellulosic biomass, and hydrogen from natural gas, electrolysis and biomass. The impact of grid electricity generation mixes on WTW energy use and GHG emissions of PHEVs was also investigated.

**A Comparison of the Full Costs of Ethanol and Gasoline**

Hill, J. D., University of Minnesota

Originally presented on: 8/11/2009 14:35:00 - 15:00:00

Many alternative transportation fuels are technologically feasible, but whether they are desirable replacements for petroleum from a societal standpoint requires that their full cost of production and use be considered, including both direct and external costs. Here the life-cycle climate change and health effects of greenhouse gas (GHG) and fine particulate matter (PM2.5) emissions from gasoline, corn ethanol, and cellulosic ethanol are quantified and monetized. Considered are multiple biorefinery heat and power sources (natural gas, corn stover, and coal), technology improvements in corn ethanol production, and multiple feedstocks for cellulosic ethanol (prairie biomass, Miscanthus, corn stover, and switchgrass). The development of a geographically-explicit spatial life cycle assessment (LCA) to estimate the human health impact of PM2.5 is also described. External air pollution costs are then compared to direct production costs, accounting for variability in input prices as experienced in a growing biofuels industry. The importance of drawing such comparisons in driving energy policy toward sustainable alternatives is emphasized.

# Trends in Sustainable Transportation


## Authors:

Tamor, M., Ford Research & Advanced Engineering



It is easily forgotten that a century ago electric vehicles had a dominant market share over steam and internal combustion automobiles. Quiet, clean, reliable and easy to use - the 'ladies car' - the EV suffered range and performance limitations, and was quickly overwhelmed in the market by rapidly improving gasoline vehicles. Driven in large part by the quest for improved fuel efficiency and emissions - and accelerated by design opportunities and attractive customer features - the re-electrification of the motor vehicle is progressing rapidly. Electrification ranges from the mundane, such as small motors, pumps and comfort features in conventional vehicles, through a spectrum of hybrid systems (that use a small, high-power battery and electric motors to achieve a dramatic improvement in fuel economy) to a proliferation of new all-electric vehicle concepts. At the same time, our public dialog often seems to accept as fact that automobiles are responsible for a large share of GHG emissions and are far less efficient than other transportation modes, that plug-in hybrids (equipped with a larger battery and external charging capability) and pure EVs will reduce greenhouse gas emissions while reducing the cost of personal transportation, that deployment of EVs and PHEVs will somehow accelerate the introduction of clean electric generation, and that battery technology has progressed to the point that a fully capable EV is feasible. Intended to promote rational discussion rather than particular solutions, this presentation combines first-principles considerations of fuel and energy storage, publicly-available energy data and results of our own vehicle studies to test these assumptions. It is found that an advanced hybrid electric vehicle is a surprisingly efficient transportation device, and with growing availability of renewable fuel it is possible to achieve sustainable personal transportation without a dramatic sacrifice of functionality. It is also suggested that public policy must be carefully designed to assure alignment between the goals of environmental responsibility and energy independence.

**Predicting Efficiency of Solar Powered Hydrogen Generation Using Photovoltaic Electrolysis Devices**

Gibson, T. L., General Motors R&D Center
Kelly, N. A., General Motors R&D Center
Originally presented on: 8/11/2009 15:45:00 - 16:10:00

Hydrogen fuel for fuel cell vehicles can be produced by using solar electric energy from photovoltaic (PV) modules for the electrolysis of water without emitting carbon dioxide or requiring fossil fuels. In the past, this renewable means of hydrogen production has suffered from low efficiency (2-6%), which increased the area of the PV array required and therefore, the cost of generating hydrogen. A comprehensive mathematical model was developed that can predict the efficiency of a PV-electrolyzer combination based on operating parameters including voltage, current, temperature, and gas output pressure. This model has been used to design optimized PV-electrolyzer systems with maximum solar energy to hydrogen efficiency. In this research, the electrical efficiency of the PV-electrolysis system was increased by matching the maximum power output and voltage of the photovoltaic array to the operating voltage of a proton exchange membrane (PEM) electrolyzer, and optimizing the effects of electrolyzer operating current, temperature, and pressure. The operating temperature of the PV modules was also an important factor studied in this research to increase efficiency. The optimized PV-electrolysis system increased the hydrogen generation efficiency to 12.4% for a solar powered PV-PEM electrolyzer that could potentially supply enough hydrogen to operate a fuel cell vehicle.

**Third-Generation Cellulosic Biofuels: Sustainable, Efficient, Cost-Effective**

Eggeman, T., ZeaChem Inc.

Originally presented on: 8/11/2009 16:10:00 - 16:35:00

Increased emphasis is being placed on renewable, sustainable, domestic forms of energy, including transportation fuels, at national, state and local levels. Successful deployment of alternative solutions to petroleum-based liquid fuels for the transportation industry is dependent upon technical, economic and environmental factors.

ZeaChem Inc. has developed a hybrid cellulose-based biorefinery platform capable of producing third-generation biofuels and biochemicals. This new approach to cellulosic biofuels provides significant technological, economic and environmental advantages compared to other, current biofuels production processes.

ZeaChem's third-generation cellulosic fuel and chemical technology is differentiated from first and second generation technology by a number of key elements:

1. The hybrid process ? incorporating biochemical and thermochemical processes ? delivers the highest theoretical yield of any biorefinery technology, a 40% advantage.

2. Use of proven technology and process components and a naturally occurring organism.

3. Feedstock flexibility allows for geographic diversity of biorefinery locations.

4. Product flexibility of cellulosic biofuels and biochemicals production to meet market demand for renewable products.

**Technology**

ZeaChem's unique hybrid process combines the best of biochemical and thermochemical platforms. Utilizing a naturally occurring acetogen, found for example in termites, ZeaChem's process fully utilizes all of the available carbon from the cellulose and hemi-cellulose fractions of the feedstock, while adding hydrogen from the lignin fraction to create the highest net conversion efficiency of any known process. By efficiently extracting the most energy possible from sustainable energy crops, ZeaChem significantly increases output while reducing both production costs and environmental impacts.

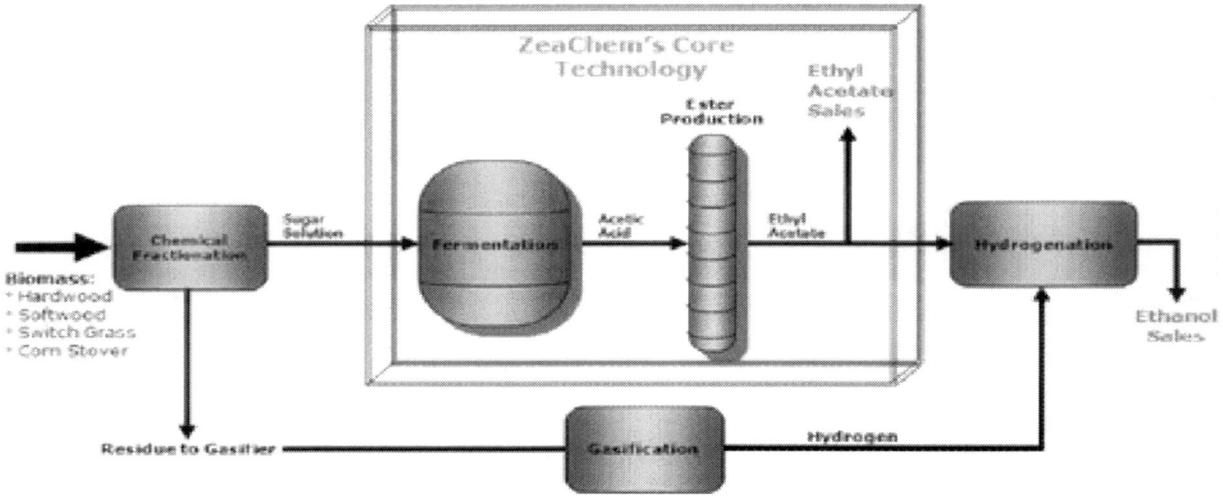


As the diagram above shows, after fractionating the biomass, the sugar stream (both xylose [$C_5$] and glucose [$C_6$]) are sent to fermentation where an acetogenic process is utilized to ferment the sugars to acetic acid without $CO_2$ as a by-product. In comparison, traditional yeast fermentation creates one molecule of $CO_2$ for every molecule of ethanol. Thus the carbon efficiency of the ZeaChem fermentation process is nearly 100% vs. 67% for yeast. The acetic acid is converted to an ester which can then be reacted with hydrogen to make ethanol. To get the hydrogen necessary to convert the ester to ethanol, ZeaChem takes the lignin residue from the fractionation process and gasifies it to create a hydrogen-rich syngas stream. The hydrogen is separated from the syngas and used for ester hydrogenation and the remainder of the syngas is burned to create steam and power for the process.

The net effect of combining the two processes is that about 2/3 of the energy in the ethanol comes from the sugar stream and 1/3 comes from the lignin steam in the form of hydrogen. The process is nearly balanced with the necessary steam and power generated from the non-hydrogen portion of the syngas stream.

For every bone dry ton (BDT) of feedstock, ZeaChem's technology will produce a theoretical maximum of 160 gallons of ethanol, with an expected yield of approximately 135 gallons per BDT (gal/BDT). Accounting for yield per acre, this is five times more than corn based ethanol processes and about 1.5 times more than other cellulosic processes, either biological (enzymatic) processes or thermochemical (gasification) processes. ZeaChem's high-yield process gives it sustainable competitive advantage in both economics and environmental impact.

**Economic**

Given its 40% yield advantage, ZeaChem's process has the lowest operating (<$1.00) and capital cost (<$3.00) per gallon. ZeaChem's advantage in operating costs puts its process at a competitive advantage to both corn based ethanol producers and cellulosic ethanol competitors. Assuming a yield of 15 BDT/acre, a biomass farm with an approximate 5 mile radius could support a ZeaChem biorefinery with capacity of 100 million gallons per year (MM GPY). This translates into net land productivity rates nearly five times higher than existing approaches, a substantial environment improvement and benefit that also translates into significant cost advantages.

| | Farm Yield | Factory Yield | Auto Efficiency | Land Productivity |
|---|---|---|---|---|
| 1st Generation - Corn Ethanol | 150 Bu/ac/yr | 2.7 Gal/bu | 14 mi/Gal EtOH | 5,670 mi/ac/yr |
| 2nd Generation ? Cellulosic Ethanol | 7.5 BDT/ac/yr | 90 Gal/BDT | 14 mi/Gal EtOH | 9,450 mi/ac/yr |
| 3rd Generation ? Advanced Cellulosic Ethanol (ZeaChem Inc.) | 15 BDT/ac/yr | 135 Gal/BDT | 14 mi/Gal EtOH | 28,350 mi/ac/yr |
| 3rd Generation + Auto Efficiency | 15 BDT/ac/yr | 135 Gal/BDT | 25 mi/Gal EtOH | 50,625 mi/ac/yr |

ZeaChem's technology has the ability to produce a wide range of biofuels and biochemicals. This flexibility allows for ZeaChem to provide the most economic, renewable products to the marketplace. Currently, ZeaChem is focusing its deployment efforts on two carbon ($C_2$) products, including ethanol. Future technology deployment will expand into other carbon chain product groups and include biofuels such as butanol, hexanol and hexene production.

| **ZeaChem Carbon Chain Product Groups** | | | |
|---|---|---|---|
| **C2 Chain** | **C3 Chain** | **C4 Chain** | **C6 Chain** |
| Acetic Acid<br>Ethyl Acetate<br>Ethanol<br>Ethylene<br>Ethylene Glycol | Lactic Acid<br>? Propylene Glycol<br>? Acrylic Acid & Esters<br>Propionic Acid<br>? Propylene<br>? Methacrylic Acid & Esters | Butanol | Hexanol<br>Hexene |

**Environmental**

The ZeaChem process results in a 40% yield advantage of ethanol per ton of feedstock compared to other conversion technologies because the process efficiently utilizes all fractions of the feedstock.

A third-party lifecycle analysis found that ZeaChem cellulosic ethanol has an advantage over traditional ethanol with its high biorefinery yields and low fossil fuel usage. The ZeaChem process was found to be a highly efficient, low net GHG emissions method of producing ethanol and was found to have an advantage in both net energy value (NEV), and net energy ratio (NER). (NEV is the energy content of the ethanol minus the fossil fuel energy input into the process of making the ethanol. NER is the ratio of output energy of ethanol to input energy of fossil fuels.)

The analysis was based on a conservative dedicated farm producing 10 BDT/acre/year with $N^{th}$ plant ethanol yield of 135 gal/BDT in comparison with published corn ethanol data from the USDA. As shown, the ZeaChem process has a very high NEV when compared to corn. The analysis also determined that ZeaChem ethanol has about eight times lower fossil fuel $CO_2$ emissions than traditional corn ethanol and about eleven times fewer emissions than conventional gasoline. Likewise, ZeaChem's high yield will provide a competitive advantage with respect to both NEV and $CO_2$ over any cellulosic competitor.

| | **ZeaChem Ethanol** | **Traditional Corn Ethanol** | **Conventional Gasoline** |
|---|---|---|---|
| Crop Yield (tons/acre) | 10* | 3.5 | N/A |
| Biorefinery Yield (gallon/ton) | 135 | 93.48 | N/A |
| Net Energy Value (BTU/gal) | 70,032 | 24,770 | N/A |
| Net Energy Ratio | 12.5 | 1.4 | N/A |
| Pounds of $CO_{2e}$ per BTU (lbs/MMBTU) | 19.12 | 165.04 | 218.38 |

Utilizing dedicated energy crops for biofuels production is a key component of ZeaChem's biofuels strategy from both an economic and environmental standpoint. Sustainable dedicated energy crops are cost effective due to the ability to lock in long-term contracts, have a low-carbon footprint based on maximizing yield per acre and harvesting techniques, and when planned in coordination with a biorefinery, offer significant transportation and storage logistics simplification.

ZeaChem has partnered with GreenWood Resources (GWR), a leading hybrid poplar tree farm operator. Engineering is currently underway for ZeaChem's first biorefinery, which is proposed to be built near GWR's existing tree farm in Boardman, Oregon.

ZeaChem believes that third-generation biofuels that balance technological, economic and environmental impacts, offer a sustainable solution to the transportation industry.

**Ethanol Industry in the United States and in Brazil: Sustainability Considerations**

Chum, H., National Renewable Energy Laboratory
Seabra, J. E., National Renewable Energy Laboratory
Hill, J., University of Minnesota
Tiffany, D., University of Minnesota
Macedo, I. C., Unicamp
Originally presented on: 8/11/2009 16:35:00 - 17:00:00

In the past thirty years the ethanol industry in the United States (US with corn) and the Brazilian (sugarcane) made significant technological improvements both in the feedstock production, conversion technologies, and their systems' integration. , Improvements in energy efficiency occurred in both countries with the learning experience and increasingly in Brazil with the production of electricity for sale to the grid in long term contracts. Also improved in both countries were conversion yields and system efficiency and production practices in the field to enhanced production capacity. Greenhouse gas emissions (GHG) have decreased following these improvements with Brazilian ethanol GHG emissions being reduced to a greater degree as more electricity is generated from bagasse and high amounts of thrash from mechanical harvest of sugarcane. Corn ethanol production using a biomass residue (e.g., corn stover) as a source of heat, power and export electricity could also reduce GHG of corn systems further (see figure 1). The semi-perennial culture growing in a tropical environment has a superior performance over an annual crop grown in temperate zones. However, this disadvantage for corn is projected to decrease if lignocellulosic biomass ethanol becomes commercially viable in the US.

Figure1. Lifecycle Greenhouse Gas Emissions Associated with Different Feeds/Processes

Acknowledgments: This work was sponsored by the U.S. Department of Energy, Office of the Biomass Program and by the Brazilian Council for Scientific Research (CNPq) as part of the bilateral collaboration between the governments of the two countries on advanced biofuels

# Second Thoughts On Preferred End-of-Life Treatment Strategies for Consumer Products


## Authors:

Dewulf, J., Ghent University
Duflou, J. R., University of Kentucky
Ostaeyen, J. V., Katholieke Universiteit Leuven



Traditionally eco-design has been steered by an implicit hierarchy of preference with respect to the end-of-life options for products of which the total life cycle impact is to be minimized. In this context life time extension is typically preferred over disassembly for reuse of components, which in its turn is preferred over material recycling. However, this priority hierarchy is often too simplistic to accept it as a general applicable guideline: both ecological and economic considerations can make life time extension and/or the reuse of components non-favourable strategies in cases where product performance and resource efficiency may evolve rapidly as result of continuous innovation. Furthermore, where ecological indicators might confirm the suitability of a life time extension strategy at a component level, economic constraints often make such scenarios infeasible. De facto today few disassembly activities prove to be economically viable. However, a number of newly emerging technologies and business strategies can be observed that could indicate a reversal of the trend to abandon the higher priority end-of-life treatment methods for manufactured goods. Based on extensive case study driven research, successful business models were revealed that improve the economic viability of systematic product reuse, refurbishment or disassembly in function of component reuse. Where material recycling proves to be the only realistic scenario, newly emerging self-disassembly techniques could help to improve the feasibility of pure material fraction separation before shredding is applied.


Keywords: End-of-life treatment, eco-design, life time extension, reuse, self-disassembly, product service systems

**Sustainable Production**

Nasr, N., Rochester Institute of Technology
Originally presented on: 8/11/2009 14:10:00 - 14:35:00

The growing demand for new products coupled with the rise in consumption in developing economies are straining the environment around the world. Industry is becoming aware that, to meet today's demanding environmental standards, the life-cycle impacts of a production system must be understood and optimized. In response, many designers are focusing on the environmental impacts of products at every stage of their life-cycle. This presentation will discuss the latest developments related to sustainable production and highlight both worldwide and national initiatives to encourage sustainable production. It will also present a systemic approach to development of metrics for sustainable production.

**Improving the Sustainability of Metal On Metal Hip Implants Via Better Machining**

Dillon, O., University of Kentucky
Deshpande, A., University of Kentucky
Puleo, D. A., University of Kentucky
Pienkowski, D., University of Kentucky
Jawahir, I., University of Kentucky
Originally presented on: 8/11/2009 14:35:00 - 15:00:00

This is a report on ?work still in progress? in which metal on metal (MoM) hip implants are used as an example of manufacturing for better product sustainability. Debris generated during normal use (wear and fatigue) of the hip implants is the main cause of prosthetic replacement. This debris (some less than 30 nm in size) elicits an inflammatory reaction which can lead to pain and failure of the implant that requires revision surgery. Hence, the sustainability issue is to improve the quality of life by manufacturing prostheses that generate less debris and last longer with reduced pain..

Fortunately, the recent literature contains a detailed description Co-Cr-Mo alloy ball and head prostheses that were removed from patients along with a study of their properties. The metallurgical microstructure of the material adjacent to the articulating surface adjusts with use. This region has several layers of differing characteristics. In the used implants, a thin layer near the interface seems to be composed of nanograin sized material, which was not the case when they were implanted. Such small grained materials are hard but brittle. Hence, the sustainable solution is to induce larger compressive residual stresses in the surface during the manufacturing (machining) process.

In order to study the implant debris generation, a purpose-built biaxial pin on disk (PoD) testing system was used to study wear of Co-Cr-Mo cylindrical specimens (0.3125 inch diameter by 1 inch long) while immersed in a simulated biofluid environment. The discs were prepared from 2 inch bar stock of the same alloy. The average sliding speed between the pin and disk was 50 mm/s. Three pins were used in each experiment for statistical purposes. Mass loss, debris generation and change in surface roughness were measured at 100,000 cycle increments up to 500,000 cycles. Preliminary data shows that the cutting tool nose radius and feed rate were the major factors affecting pin wear rate. Micro-hardness and residual stresses are also measured. Pin wear rate was found to be proportional to bearing surface microhardness. These results demonstrate that machining

conditions can be selected which will induce compressive residual stresses in C0-Cr-Mo alloy MoM hip implants.

**Sustainability Challenges and Opportunities in Nanoelectronics Manufacturing -**
Shadman, F., University of Arizona
Originally presented on: 8/11/2009 15:00:00 - 15:25:00

Manufacturing in nano-scale is not a simple extrapolation of the current and conventional manufacturing practices. The shift to nano scale creates both major issues as well as significant opportunities in performance, cost, and environmental impact for electronics industry. This presentation focuses on both the positive and the negative environmental impacts of this inevitable paradigm change. The inter-relationship between the three sustainability factors (performance, cost, and environmental impact) for future electronics manufacturing will be analyzed. In particular, the challenges related to resource requirements and utilization (water, energy, and chemicals) will be discussed. Examples of unique opportunities for environmental gain, that nano-scale manufacturing can provide, will be presented.

**Excellent Product Stewardship and Sustainable Use of Flame Retardants**

Landry, S., Albemarle Corporation
Scherrer, S., Chemtura Corporation
Tenney, J., Albemarle Corporation
Originally presented on: 8/11/2009 15:45:00 - 16:10:00

The use of flame retardants reduces ignitability, delays the spread of fires and delays the time of flashover in order to give people more time to escape, and facilitate the intervention by firefighters. The ultimate purpose of their use is to save lives, reduce injury, reduce destruction of property, and reduce local pollutants that result from fires.

Industry is responding to the challenge to ensure that flame retardants are sustainable. As part of the commitment to responsible production and use of flame retardants, the Bromine Science and Environmental Forum (BSEF) has launched an emissions reduction program. By focusing on excellent Product Stewardship, the responsible management of the health, safety, and environmental aspects of flame retardants throughout the life cycle are addressed. This program is called the Voluntary Emissions Control Action Program (VECAPTM). This unique emissions control program was developed to address emissions of chemicals, with flame retardants being the first to utilize the program.

Reducing or eliminating emissions of chemicals to air and water can help to insure the safe, environmentally responsible, and sustainable use of these products. The flame retardant industry is firmly committed to providing needed resources and leadership to help provide safe and sustainable products that provide a valuable service to society. This paper addresses the details and progress of VECAP and potential benefits to the plast

**Plastics Products Exemplifying Sustainable Development: Life Cycle Inventory of Metallized OPP Film in Packaging Applications**

Sahnoune, A., ExxonMobil Chemical Co.
Johnson, E., ExxonMobil Chemical Co.
Originally presented on: 8/11/2009 16:10:00 - 16:35:00

Over the past few years sustainability and sustainable development have become major areas of concern for the public, policy makers, and companies alike. In the area of plastics, the environmental impacts of products derived from these materials, such as grocery bags and packaging, have been and continue to be the subject of a number of studies. Plastics products are present in every aspect of our daily life, from use in medical and space applications to toys and grocery bags. Their rapid adoption over the years stems from their reliability, ease of manufacturing, and affordability. However, with the heightened awareness about the limited energy resources, the plastics from conventional sources such as natural gas (NG) and oil are sometimes perceived to negatively impact the environment and sustainability in general. Through a detailed life cycle inventory (LCI) analysis of metallized oriented polypropylene films (mOPP), we show that plastics can have many positive attributes and in many cases would be the preferred solution from a sustainability perspective.

Indeed, the LCI indicates that the use of mOPP saves materials, consumes less energy and emits less green house gasses (GHG) during the life cycle of the product from cradle to grave. The talk will also highlight other examples of plastics products and discuss their respective contributions to sustainability.

**Methodology of Product Disassemblability Analysis**

Afrinaldi, F., Andalas University
Mat Saman, M. Z., Universiti Teknologi Malaysia
Shaharoun, A. M., Universiti Teknologi Malaysia
Originally presented on: 8/11/2009 16:35:00 - 17:00:00

Nowadays, most countries around the world have set legislations which are planned to force manufacturers to reuse, recycle, recover and remanufacture their products at the end of their life. However, before end-of-life products can be recycled, the end-of-life disassembly needs to be in place. This entails large amounts of capital expenditure. Most companies would not like to even consider disassembling and remanufacturing unless capital costs are justified and financial gains assured. Besides that, in general, the designers also do not have experience in disassembling and recycling to determine impact of various design aspects on difficulty and effort at the disassembly stage. Based on those, there is a strong need for a methodology to evaluate the disassemblability and determine the end-of-life options of the end-of-life products. It is expected to provide assistances in making decision at the early stage of product design. The end-of-life option determination and the disassemblability evaluation will show how economically efficient is it to disassemble a product and check the opportunity of a component to be recycled. This paper proposes a methodology to fulfill those needs. The proposed methodology is divided into five stages: (1) product definition; (2) end-of-life value calculation; (3) disassemblability evaluation; (4) recycling rate calculation; and (5) end-of-life value, disassemblability and recycling rate report.

Keywords: disassembly, disassemblability, disassembly seaquence, end-of-life, recycling

**Fuzzy Approach to the Assessment of Sustainability Indices**

Imanov, G., Institute of Cybernetics of the National Academy of Azerbaijan Republic

This paper considers the sustainability of development level and environmental sustainability indices in the States. For the assessment of these indices we propose application of fuzzy logic inference method. Problem solution is demonstrated on the basis of International and Azerbaijani statistical data.

**Symbiotic Environs**

Mangone, G., University of Virginia

The design approach of integrating bioorganic and natural processes with the infrastructural functions of the built environment sustains a healthier environment that educates inhabitants about the quantity of their resource consumption and develops an understanding and appreciation for the ecological processes necessary for sustaining life through a multi-sensory engaging constructed landscape. The resultant built environment operates with significantly less resource demand and enhances the natural environment through the addition of nutrients, rather than the current practices of resource depletion and pollution. This task will be pursued through the programmatic framework of an innovative office building situated within a residential network in Tacoma Park, Washington D.C, and follow the innovative Cisco Telepresence office building concept. The focus will be on the design, integration, and implementation of the program spaces and water distribution, collection, filtration, and storage infrastructure with regards to providing maximum thermally optimal occupied spaces throughout the year. This is achieved through the potential for interrelation of a cave ecosystem with a greenhouse, supplemented with gradient space. The primary focus is the relationship of these systems to the occupancy of the building, evaluating the effects of these factors on each other, and determining opportunities to propose new types and levels of interaction and relationship between these systems. Additional facets which were undertaken were agricultural production, passive energy generation, minimization of the embodied energy footprint of materials and structure, and collecting, filtering, and storing surrounding community greywater and storm water. Throughout the course of the research it was determined that the integration of water collection, storage, filtration, and distribution did not have a direct impact on the form or organization of the building. This system can be integrated in a plethora of ways to various spatial/programmatic strategies. As well, both the greenhouse and cave spaces were optimal for different programs, notably cave spaces can be utilized for maximum contrast environments for digital projection, work, and conference environments; while naturally daylit greenhouse and gradient spaces provide optimal collaboration and lounge spaces. The overall conclusion was that varied spatial environments in regards to thermal, visual, and acoustic qualities provide a more cohesive, adaptable, and efficient work environment that allows for a broader and more optimal range of work, while performing at maximum efficiency and serving as a community and environmental asset. This develops a gradient between built and natural environments, through co-habitation and blurring of these environs.

**Using off-Peak Wind to Recycle CO2 Into Transportation Fuels**

Doty, D., Doty Scientific
Holte, L., Doty Scientific
Shevgoor, S., Doty Scientific

The absence of satisfactory energy-storage solutions has recently led to the increasing availability of negative-priced off-peak grid energy in areas of substantial build-up of wind energy. Putting the excess off-peak wind energy into stable synthetic liquid fuels should permit complete grid stabilization and at the same time address the need for carbon-neutral fuels for transportation.

It has long been accepted that it should theoretically be possible to convert CO2 and water into standard liquid hydrocarbon fuels at high efficiency. The problem has been that prior proposals for doing this conversion have had efficiencies of only 25% to 35%. That is, the chemical energy in the liquid fuels produced (gasoline, ethanol, etc.) would be less than the 30% of the input energy required. Our work shows that a factor-of-two gain in efficiency should be possible.

Doty Energy is developing novel processes that will allow the production of fully carbon-neutral ethanol, gasoline, jet fuel, diesel, and most plastics from waste CO2 and off-peak wind or other low-carbon energy at high efficiency and at prices that should soon be competitive with fossil-derived products. Converting CO2 into fuels can eliminate the need for CO2 sequestration and reduce global CO2 emissions by 40%.

The US transportation-fuel market is about $1T, and it is based mostly on imported oil. A better alternative for future transportation fuels is needed than those that have previously been advocated ? such as biofuels, hydrogen, and methane. It is in our economic and security interests to produce carbon-neutral transportation fuels domestically at the scale of hundreds of billions of gallons annually. Most biofuels are only 5% to 25% carbon neutral; and hydrogen has daunting cost challenges with respect to distribution, storage, and end use. Windfuels will be over 85% carbon neutral and will flow seamlessly within our current infrastructure.

The combination of the major technical advances we have simulated over the past two years should permit conversion of CO2 to fuels to be done at up to 60% efficiency, which is about twice what was expected by most researchers just three years ago. When the input energy is from off-peak wind or nuclear and reasonable consideration is included for climate benefit, WindFuels will sometimes compete even when petroleum is $45/bbl, and most experts are predicting oil will stay over $140/bbl after 2014. Our analysis shows windfuels production per gross land area in good wind regions should exceed biofuels production density in productive farming areas by factors of 4 to 20. Moreover, 99.9% of the land required for the wind farms is generally available for dual use. Switching 70% of global transportation fuels from petroleum to WindFuels should be possible over the next 35 years.

The WindFuels production system (disclosed in detail in pending patents) will be presented. It is based largely on the commercially proven technologies of wind energy, water electrolysis, and Fischer Tropsch (FT) chemistry. Wind energy is used to split water into hydrogen and oxygen. Some of the hydrogen is used in a process, the so-called reverse water gas shift (RWGS) reaction, that reduces CO2 to carbon monoxide (CO) and water. The CO and the balance of the hydrogen are fed into an FT reactor similar to that commonly used to produce fuels and chemicals from coal or natural gas.

Windfuels are truly sustainable. The needed CO2 could ultimately come from biofuels refineries, cement factories, ore refining, and possibly even the atmosphere. The water requirements are an order of magnitude less than for biofuels, and the wind will always blow.

The biggest part of the increase in our projected efficiency comes from improved separations processes in the fully recycled FT loop. Conventional processes for separation of CO2 from the other syngas components have typically required over 6 MJ/kg of CO2. Our simulations indicate a high-pressure cryogenic process can achieve sufficient separation in the FT recycle loop (under 15 molar-%) at under 0.8 MJ/kg. The next largest gain is expected from an order-of-magnitude advance in cost-effectiveness of gas-to-gas recuperation, which is expected to make 97% effectiveness practical where 75% was previously practical. The third largest gain may be from an optimized RWGS process which promises over 90% efficiency (implying under 1.55 MJ per kg of CO) compared to about 50% for prior demonstrations (about 2.8 MJ/kg-CO). Another significant gain comes from a novel thermodynamic cycle for conversion of the waste heat from the electrolyzer and the FT reactor. A novel approach to RFTS system integration and optimization leads to additional efficiency gains. All the processes have been simulated in detail, and key experiments will soon be carried out to help optimize process conditions.

References

1. FD Doty et al, ?WindFuels ? Competitive Fuels from Wind Energy and Waste CO2?, presented at AIChE annual meeting, Philadelphia, 2008 http://windfuels.com/PDFs/WindFuels_Sci_Engr_Basics_doc1.pdf.

2. SE Park, JS Chang, KW Lee editors, Carbon Dioxide Utilization for Global Sustainability, Studies in Surface Science and Catalysis 153?; OS Joo, KD Jung and J Yonsoo, ?CAMERE Process for methanol synthesis from CO2 hydrogenation?, pp 67-72, Elsevier, 2004.

3. Mark Z Jacobson, ?Review of Solutions to Global Warming, Air Pollutions, and Energy Security?, Energy Environ. Sci., 2009.

4. P.L. Spath, D.C. Dayton, (NREL/TP-510-34929, 2003).

5. O.S. Joo, K.D. Jung J. Yonsoo in. Carbon Dioxide Utilization for Global Sustainability, Studies in Surface Science and Catalysis 153, S.E. Park, J.S. Chang, K.W.

Lee, Eds., pp 67-72, Elsevier, 2004.

6. M Xiang, D Li, H Qi, W Li, B Zhong, Y Sun, ?Mixed alcohols synthesis from CO hydrogenation over K-promoted β-$Mo_2C$ catalysts?, Fuel 86, 1298-1303, 2007.

7. R. Zubrin, B. Frankie, T. Kito, AIAA 97-2767, (1997).

8. S Phillips, A Aden, J Jechura, and D Dayton, ?Thermochemical Ethanol via Indirect Gasification and Mixed Alcohol Synthesis...?, NREL/TP-510-41168, 2007.

9. FD Doty and S Shevgoor, ?Compact, High-Effectiveness, Gas-to-gas Compound Recuperator with Liquid Intermediary (CRLI)?, HT2009-88372, ASME Joint Conferences, San Francisco, 2009.

10. FD Doty and S Shevgoor, ?A Dual-source Organic Rankine Cycle (DORC) ...? ES2009-90220, ASME Joint Conferences, San Francisco, 2009.

11. GN Doty, FD Doty, LL Holte, D McCree, S Shevgoor, ?...Efficiently Recycling CO2 into Transportation Fuels ? Driving the Off-peak Wind Market?, Proceed. WindPower 2009, paper 175, Chicago, 2009.

**PRODUCTION of Energy, Biofuels and Potable Water with AN Integrated Resource Oasis**

Moolick, R. T., Arizona State University

While no single technology is the complete solution to resource shortages, an integrated system of technologies can significantly contribute to shifting the resource balance towards sustainability. Proposed is a ?resource oasis? capable of producing electricity, biofuels, potable water and other commercially attractive products from renewable resources. The ?oasis? uses multiple energy inputs such as wind, solar and tidal energies to produce electricity, fuels, potable water and other useful products with minimal environmental impacts. The ?oasis? is a modular floating platform allowing the efficient capture of solar energy through photovoltaic and biological systems. Omni-directional wind turbines and open-water current turbines supplement the energy inputs. Taking in seawater, the ?oasis? produces ethanol, biodiesel, high-protein feed, hydrogen, and potable water. The modules can be readily linked together to increase outputs. The modules can be anchored offshore transferring products onshore through pipelines, or anchored in deep water transferring products via ship.

**Optimizing the Solar Photovoltaic Energy Capture On Sunny and Cloudy Days Using a Solar Tracking System**

Kelly, N. A., General Motors R&D Center
Gibson, T. L., General Motors R&D Center

Solar energy is a clean, renewable way to provide electrical power, as well as to power the future hydrogen economy. We envision using solar-powered photovoltaic (PV) devices to drive the electrolysis of water to make hydrogen in both centralized and home fueling systems for fuel-cell electric vehicles (FCEVs). In this way the solar energy is stored in the hydrogen. At the GM R&D Center, we have investigated ways to optimize the solar to hydrogen process by photoelectrochemical and photovoltaic-electrolyzer (PV-electrolyzer) systems. By improving the efficiency of the solar-water-splitting process, we can help to bring a solar-powered hydrogen home fueler for closer to reality. Using renewably-generated hydrogen to power future FCEVs can eliminate air emissions from vehicles and contribute to the diversity of the possible sources of hydrogen. PV-electrolyzer systems have only been developed for the purposes of demonstration and proof of concept due to their high system costs, system complexity, and, in some cases, low system efficiencies for generating the high-pressure hydrogen needed for hydrogen storage. Although it is feasible to optimize the efficiency of PV-electrolyzer systems for making hydrogen, there is also a need to improve the individual solar and electrolysis systems. We have determined a method in which a 2-axis solar tracking system is optimized to capture the maximum amount of solar energy on both sunny and cloudy days. Previously, solar tracking systems have only been optimized for sunny days, when direct sunshine that accounts for 85-90% of the total solar radiation, predominates. However, improving the harvesting of solar energy on cloudy days is important to using solar energy on a daily basis for fueling FCEVs because it improves the hydrogen yield on the days with the lowest hydrogen generation, which in turn reduces the system size and cost.

In order to design an optimal solar tracking system for capturing solar energy we have analyzed an extensive set of measurements of the solar irradiance obtained using four identical solar arrays and associated solar sensors (collectively called solar collectors) with different tilt angles relative to the earth's surface. The measurements included a variety of ambient conditions including different seasons and both cloudy and cloud-free conditions. One solar array (and its associated solar sensor), for a period around solar noon, was always approximately pointed directly toward the sun (DTS); it captured the direct component of the solar radiation that predominates on sunny days. On sunny days, solar arrays and sensors with a DTS configuration captured up to twice as much solar energy as solar collectors with a horizontal (H) orientation in which the flat surface of the array was tilted toward the zenith. This finding was in agreement with the cosine response law for flat-surfaced solar collectors. Another set of solar collectors always had an H orientation, and this best captured the diffuse component of the solar radiation that predominates on cloudy days. During cloudy periods, we found that an H configuration increased the solar energy capture by nearly 50% compared to a DTS configuration during the same period. A simple model for diffuse radiation, referred to as the Isotropic Diffuse Model, agrees qualitatively with the angular dependence that we measure for the horizontal/2-axis tracking irradiance ratio on cloudy days. However, it predicts a lesser dependence for the magnitude of this ratio than we measured. We propose three embodiments for an optimized 2-axis tracking system that uses simple algorithms and sensors to determine when it is cloudy, and then orients the solar modules to an H configuration for maximum solar system performance the during periods of cloud cover.

The importance of improving the solar energy output on cloudy days is illustrated in the following scenario for a home hydrogen fueling system. Assume that a solar-powered hydrogen fueling system that is designed to produce, on average, sufficient solar energy to generate 0.5 kg of hydrogen per day via water electrolysis. This amount of hydrogen would be sufficient to for an FCEV on an average daily commute of 30 miles. Typically, solar systems are sized to produce the desired amount of energy on an average solar day for the location of interest. The size (area) of the solar system depends many factors, including the location, the season, and the weather on a given day. The output of solar systems is highly variable; on short, cloudy winter days the system will make much less that the average amount, while even on the longest sunniest summer day it will make no more than twice the average amount. Ideally, the hydrogen storage system must be able to hold enough hydrogen to supply the vehicle for a few bad days in a row ? this is like the autonomy built into an off-grid solar system that uses batteries to store solar energy. Of course the vehicle can occasionally use the centralized hydrogen dispensing system, but this is to be avoided as much as possible to maximize the use (convenience and cost) of the solar home fueling system. Therefore, improving the solar energy capture on the worst days is important to minimize the solar and storage system size and maximize the overall use of the home fueling system.

**The Complexity of Assessing Process Sustainability for International Development Programs**

Quazi, H. A., AngelNet Energy Water & Environmental Systems

This paper will discuss the critical issues and complexity in assessing sustainability of processes for international development programs. Programs from Asia, Latin America, and North America will be discussed and an assessment methodology will be presented. These programs, funded by international financing and development agencies, include processes for manufacturng pulp & paper, chemicals & ferlizer, and petrochemicals. Some of these programs were successfully implemented.

**Study of the New Type Solvent of Cellulose**

Zhang, X., Shijiazhuang University
Liu, H., Shijiazhuang University
Lv, L., Shijiazhuang University

In recent decades, the development and utilization of natural polymer have been attracted increasingly because of the consumption and over-exploitation of non-renewable resources. Cellulose which consists of â-(1-4)-linked glucose repeating units is the largest renewable biological resources, it has been widely used in industrial domains such as fiber, paper, polymer, textile and food industries. However, natural cellulose is high crystalloid, strong inter- and intra-molecular hydrogen bonding caused by the hydroxyl groups and insoluble in water and common organic solvents, these properties affect their development and utilization. Therefore, the research and development of new fiber with green solvent systems, in particular, which the homogeneous reaction of cellulose derivative can occur in cellulose solvents, become a hotspot. To circumvent the use of metals in ionic fluids a variety of closed shell anions have been synthesized and these have been the main focus of ionic liquids research over the past decade. Most studies have centered on imidazolium based cations with a range of anions including BF4-, PF6-, and (CF3SO2)2N-. While these solvents show excellent solubility for a wide range of solutes and unusual miscibility with molecular liquids their widespread application will be limited by toxicological, economic and purity issues. Since none of these ionic liquids have yet been registered their general use as solvents may be restricted. Recently, we design and prepare some solvent composed of hydrogen bond, such as ionic liquid, deep eutectic solvent (DES), in which insoluble material can be dissolve. some eutectic mixtures of caprolactam, acetamide and urea were prepared. The melting point of the eutectic mixture is depressed. For example, the melting point of caprolactam/acetamide mixture in 1:1 mole ratio is 18°C, which is considerably lower than that of either of the compounds (mp caprolactam = 68~70 °C and acetamide = 78~81 °C). The significant depression of the melting point arises from an interaction between two organic molecules by hydrogen bonding. This type DES allows the mixture to be used as an ambient temperature solvent, but the solubility of cellulose in prepared DES is small. In this contribution, a new type solvent of glycerin / H2SO4 for the dissolution of cellulose is studied specially. The solubility of cellulose increases with the mole increase of H2SO4, but cellulose is carbonized when the mole ratio is over 1:2. The solubility can be up to 15% at the mole ratio 1:2. Cellulose is insoluble in the solvent adding water, thus the solvent can be renewable by distilling the water. The dissolution mechanism of cellulose in this solvent involves oxygen of H2SO4 interact with hydrogen of glycerin in the formation of oxygen anion of glycerin. After that, the hydrogen atoms of cellulose-OH act as an electron acceptor which interacts with the oxygen anion of glycerin. Upon interaction of the cellulose-OH and glycerin oxygen anion, the oxygen and hydrogen atoms from hydroxyl groups are separated, resulting in opening of the hydrogen bonds between molecular chains of the cellulose and, finally, the cellulose dissolves. Compared to ordinary ionic liquid solvents, this solvent is cheaper to prepare easily, much less toxic and non-purified. In addition, it is environmentally friendly solvent and can be recycled. Cellulose dissolution can be realized at room temperature and atmospheric pressure.

**Desulfurization of Oil Using Ionic Liquids as Phase Transfer Catalysts**

Zhao, D., Hebei University of Science and Technology CHINA
Wang, Y., Tianjin university
Duan, E., Tianjin university

In this work, kinds of ionic liquids based on dialkylimidazolium and N-alkyl-pyridinium cations were was prepared. The ionic liquids were employed as phase-transfer catalysts (PTCs) for phase-transfer catalytic oxidation of dibenzothiophene (DBT) dissolved in n-heptane, The partition ratios of DBT to ionic liquids and n-heptane were investigated. Then H2O2(30%) was used as oxidant, DBT was oxidized without the presence of PTCs, the oxidation desulfurization rate was only about 13%. While PTCs and H2O2 were introduced together, the desulfurization rate of DBT was increased apparently. When IL([C4mpy][HSO4]) was used as PTC, and the condition were: temperature is 50°C, the time is 30min, H2O2/sulfur mol ratio (O/S) is 6, the desulfurization rate reached the maximum (92.7%). The PTC [C4mpy][HSO4] can be recycled for 4 times without significant decrease in activity.

**New Binary Mixture Ionic Liquid Based On Quaternary Ammonium Salt/Caprolactam**

Zhao, D., Hebei University of Science and Technology CHINA
Cui, H., Hebei University of Science and Technology
Duan, E., Tianjin university

In recent years, ionic liquids are attracting increased attention worldwide because they have been shown to function as novel green reaction media. As far as we know, an ionic liquid typically consists of organic nitrogen-containing heterocylic cations and inorganic anions. Here, a novel room-temperature binary mixture ionic liquid based on quaternary ammonium salt and caprolactam has been synthesized and characterized by pH, liquid-solid phase diagram, solubility, conductivity and cyclic voltammetry. This molten salt appears as a liquid at <100°C, though it is composed of two solids. The melting points of the binary ionic liquid vary with the different content of caprolactam. For example, the C6H11NO/(C4H9)4NHSO4 binary mixture ionic liquid, according to the analysis of its liquid-solid phase diagram, the eutectic point is about 28°C with the mass percentage of caprolactam 75%. This work has shown that mixtures of caprolactam with quaternary ammonium salts form low melting point binary mixture ionic liquid that have unusual solvent properties.

Key words: binary mixture, phase diagram, caprolactam

**Integrated Modeling Analysis for Sustainability Assessment and Policy Decision Making**

Shastri, Y., Energy Biosciences Institute
Diwekar, U., Vishwamitra Research Institute, Center for Uncertain Systems: Tools for Optimization and Management
Cabezas, H., U.S. Environmental Protection Agency
Williamson, J., U.S. Environmental Protection Agency, Office of Research and Development, National Risk Management Research Laboratory
Lewis, N., U.S. EPA

Successful implementation of sustainability ideas in systems management requires a basic understanding of the often nonlinear and non-intuitive relationships amongst different dimensions of sustainability, particularly the system-wide implications of human actions. This basic understanding further includes a sense of the time scale of possible future events and the limits of what is and is not likely to be possible. With this understanding, systematic approaches can then be used to develop policy guidelines for the system. This work presents an illustration of these ideas by analyzing an integrated ecological-economic-social model, which comprises various ecological (natural) and domesticated compartments representing species along with a macro-economic price setting model. The stable and qualitatively realistic model is used to analyze different relevant scenarios, which include gradual changes as well as sudden catastrophes. Apart from highlighting complex relationships within the system, it identifies potentially unsustainable future developments such as increased human per capita consumption rates. Fisher information, a measure based on information theory, is used to quantify the system-wide sustainability, and economic parameters in the system are shown to be strong predictors of future instability. Dynamic optimization is then used to develop time-dependent policy guidelines for the unsustainable scenarios using objective functions that aim to minimize fluctuations in the system's Fisher information. The results can help to identify effective policy parameters and highlight the trade-off between natural and domesticated compartments while managing such integrated systems. The results should also qualitatively guide further investigations in the area of system level studies and policy development.

**Sustainable Transportation Fuels From Atmospheric Carbon Dioxide**

Zeman, F., New York Institute of Technology

Reducing carbon dioxide (CO2) emissions from the transportation sector may be the hardest part of the carbon challenge. The sector is not amenable to the Carbon Capture and Storage (CCS) technologies as summarized in the recent IPCC special report on CCS [1]. In the U.S., the transportation sector was responsible for 27% of total GHG emissions in 2005 [2]. It is therefore responsible for over 50% of the emissions that cannot be handled by conventional CCS or conversion to nuclear or renewable power. Transportation also experienced the largest growth (24%) in direct use emissions over the last decade [2]. While there are many opportunities for continued improvements in vehicle energy efficiency, achieving the drastic reductions needed for climate protection will require a more deep-rooted approach. We propose adding carbon neutral, carbonaceous fuels as an option to the collection of technologies capable of solving the transportation challenge. The transportation sector encompasses a wide variety of applications, which suggests a portfolio approach as opposed to seeking a ?silver bullet?.

Proposed solutions for the transportation sector sometimes assume that changes will be spurred by the depletion of conventional fuels. The rapid depletion of oil reserves is indeed the subject of much speculation [3], however, there is little doubt that known fossil reserves, especially coal, are sufficient for several centuries [4]. External factors, such as energy security and population growth, or consistently high oil prices may lead to rapid development of unconventional oil sources. Sources such as oil sands, heavy crude or coal are generally more energy intensive and can increase upstream CO2 emissions anywhere from 10% to 90%, depending on the technology [5].

Given the magnitude of the carbon challenge and the limitations of existing solutions, such as corn ethanol, it is crucial to understand if other options exist for de-carbonizing the transportation sector. Here we examine several alternatives for near-zero emission transportation systems that are linked by the use of carbon-based transportation fuels. The concept is based on collecting an equal amount of CO2 from the atmosphere as is contained in the fuels. Following is a discussion of the various ways such a fuel cycle can be attained?which depend to varying degrees on CCS and/or non-fossil energy sources?and a first order comparison of the costs involved in hydrogen and electrical fuel cycles.

Fuels from the Atmosphere

Carbon neutral fuels refer to fuels containing carbon that do not produce a net increase in atmospheric CO2 content upon use. Such fuels can be achieved by capturing CO2 from the atmosphere, either as biomass or directly by industrial means (air capture). Air Capture refers to the industrial removal of CO2 directly from the atmosphere at concentrations on the order of 400 ppm [6]. The alternative is the formation of biomass via photosynthesis, the natural analogue to air capture. The ability to capture CO2 from the atmosphere opens new routes towards sustainable transportation fuels. Once captured, the CO2 may be sequestered to offset emissions from fossil transportation fuel combustion elsewhere (referred to as indirect methods). Alternatively, it may be recycled in the fuel cycle by its incorporation into a synthetic fuel, conventional biofuels, or a hybrid of the two (referred to as direct methods). The synthetic fuel pathway, for instance, depends on a source of primary energy to drive the required chemical reactions including the supply of hydrogen. As with hydrogen and electricity, these synthetic hydrocarbons are an energy carrier produced from a primary energy source such as wind, nuclear power or fossil fuels with CO¬2 capture. Unlike hydrogen and electricity, they are carbonaceous fuels that are nevertheless carbon neutral as they were derived from the atmosphere. The relationship between all of the options is presented in Figure 1.

The use of biomass systems for fuel production on an industrial scale requires a certain degree of sustainability. As noted by the IPCC [7], ?biomass production and use imply the resolution of issues relating to competition for land and food, water resources, biodiversity and socio-economic impact.? Furthermore, climate change itself may alter current patterns of water usage and arable land putting more pressure on food production and water resources, in addition to land allocated to energy production. Natural systems are very complex and subject to the law of unintended consequences. Biomass based systems may still be viable; however, they inherently contain a larger amount of risk. The risk can be summarized as the challenge of solving a climate change problem with a technology totally dependent on the climate.

**Measuring Product Design Sustainability Based On Design for Assembly and Design for Disassembly Using Life Cycle Assessment**

Memary, R., Universiti Teknologi Malaysia
Mat Saman, M. Z., Universiti Teknologi Malaysia
Bin Sharif, S., Universiti Teknologi Malaysia

Sustainable manufacturing has become imperative all over the world in the current century. Manufacturers are trying to understand the need for the responsible use and management of resources in the life cycle of a manufactured product, especially the planning for product retirement. In order to measure the design sustainability of a product, cost analysis based on design for assembly and design for disassembly cost and end of life value is used. Design for assembly is an approach to design the product in the way that has least number of parts and simplest assembly process with minimum cost required. This design for X approach has many guidelines developed mainly by quantitative evaluation method such as Boothroyd-Dewhurst Design for Assembly Method, Hitachi Assemblability Evaluation Method and Lucas Design for Assembly Methodology. In case of guidelines, reducing part count and part types, striving to eliminate adjustments, Designing parts to be self-aligning and self-locating, Ensuring adequate access and unrestricted vision, Ensuring the ease of handling of parts from bulk, Minimizing the need for reorientation during assembly, Designing parts that cannot be installed incorrectly, Maximizing part symmetry if possible or make parts obviously asymmetrical are considerable. Each of this considerations in will make different costs in each part design for assembly. On the other hand design for disassembly is an approach to find the best disassembly sequence with least time. This also consists of some guidelines like combining elements, Limiting material variability, using compatible material, Grouping harmful materials to subassemblies, Providing easy access to harmful, valuable or reusable parts, using Accessible drainage points, Using fasteners easy remove or destroy, Minimizing number of fasteners. Applying these guidelines result in different disassembly costs. In fact design for assembly severely influence design for disassembly cost which should be considered in life cycle assessment. Each option for end of life value has a cost that should justify disassembly cost; else the product will not be considered as a sustainable product. In these paper four case studies including an electric pot, a hair clipper, a speaker and a printer are analyzed based on assembly, disassembly and end of life option. If end of life option justifies the disassembly cost of a part, that part is considered a sustainable part, but steel the product may be unsustainable. If total cost analysis of parts in a product can justify product disassembly or product recycling it is considered a sustainable design. Based on the justified percent of disassembly cost by end of life value a sustainability design measurement is assigned to each product.

**Effect of Design for Assembly and Design for Disassembly On Product Recyclability**

Memary, R., Universiti Teknologi Malaysia
Mat Saman, M. Z., Universiti Teknologi Malaysia
Bin Sharif, S., Universiti Teknologi Malaysia

According to the United States Environmental Protection Agency (EPA), about 11 billion tons of industrial waste (some toxic) is generated annually in the USA.. Europe is already familiar with the problem. According to Hentschel, the volume of electronic waste in Germany exceeded 800000 tons annually in the early 1990s. Designing products for sustainable development has become a pressing concern. One way to achieve this is to undertake a life cycle assessment (LCA) for a product. LCA seeks to minimize the environmental burden arising from the manufacture, use, and eventual disposal of products. Design engineers; in particular, ought to consciously plan for product retirement. For example, should the product be discarded in a landfill, reused, or recycled in whole or in part? Unlike the first environmental revolution in the1970s, which was aimed at cleaning up hazardous waste from contaminated sites and natural resources, the Second revolution is addressing waste reduction at the source. It means bringing the problem into the product design and development process and solving it through product design tools. Between product development process stages, assembly and disassembly play key roles in supporting sustainability. Considering Design issue from manufacturers and recyclers view clarifies an obvious contradiction between their goals. A manufacturer is interested to get higher profit through decreasing the production cost and accelerating the product development process. Obviously, these aims are reachable through considering design for assembly and design for manufacturing. On the contrary a recycler is interested to enlarge the profit by considering End of Life Value and Recycling Cost. These goals are reachable only through Design for Disassembly and Design for Remanufacturing. There are many different Quantitative evaluation procedures to evaluate and improve design for assembly that mostly concentrates on evaluating the design and improving it using quantitative methods. For improving the design, eight guidelines are applied in Boothroyd-Dewhurst quantitative method including reducing part count and part types, striving to eliminate adjustments, Designing parts to be self-aligning and self-locating, Ensuring adequate access and unrestricted vision, Ensuring the ease of handling of parts from bulk, Minimizing the need for reorientations during assembly, Designing parts that cannot be installed incorrectly, Maximizing part symmetry if possible or make parts obviously asymmetrical. In addition, there are some disassembly design guidelines such as combining elements, Limiting material variability, using compatible material, Grouping harmful materials to subassemblies, Providing easy access to harmful, valuable or reusable parts, using Accessible drainage points, Using fasteners easy remove or destroy, Minimizing number of fasteners. All of assembly and disassembly guidelines have prominent effects on design which ends to suitable option for product parts end of life option. This will result on product recyclability in terms of economically justifiable recycling. In this paper based on different case studies including an electric pot, a hair clipper, a speaker and a printer different effects, that applying each guideline from assembly or disassembly view on product parts recyclability may have, are investigated

**Priorities and Financial Mechanisms of Belarus Sustainable Development**
Nesterenko, V. P., Belarusian National Technical University

Achieving sustainable development is of a global nature, remains a large task for all countries, but particularly for developing countries and countries in transitions. The strategic aim of Belarus sustainable development is defined in the National Strategy of Sustainable Development of the Republic of Belarus until 2020 (NSDS-2020) elaborated proceeding from the recommendations and principles of the UN Conference on Environment and Development (Rio de Janeiro, 1992).

The important tools of Belarus sustainable development are, namely: 1) fiscal, monetary and pricing policy; 2) facilitating market and other institutional transformations in economy; 3) resource-saving technologies and environmentally safe productions: research and innovation; 4) programs and mechanisms of production modernization, enterprise restructuring; 5) investment and structural policy; 6) regional sustainable development programs; 7) activation of foreign economic policy and international cooperation.

The key objectives of the fiscal policy are to ensure financial and social stability, create macroeconomic conditions for sustainable economic growth and improved living standards in the context of ecological concerns. Environmental protection objectives call for a set of nature conservation measures, research and development in energy efficiency and resource saving, as well as development of ecological monitoring. A significant area of budget spending is the construction of sites related to rehabilitation and recovery interventions in the Chernobyl-affected areas. A broader involvement of the private capital is envisaged in the financing of conservation and resource-saving activities.

In the forecast period, the monetary policy should serve to achieve the general strategic goals of sustainable development through introducing stability of the national monetary unit, higher reliability and liquidity of the banking system, effective and secure functioning of the system of payments. The implementation of a set of monetary and crediting activities will ensure financial stability at the macroeconomic level and attainment of parameters and scope typical of the monetary systems of the economically developed countries.

The strategic goal of institutional transformations in the Republic of Belarus is to create a system of institutions necessary to develop the market economy and able to address the issues of raising the living standards, production modernization, preservation of both productive potential of the natural complex, and security of the country. Ownership plays a decisive role in the system of institutional market changes. Ownership transformations aim to create a plural socially oriented market economy, support private initiative and enterprise, promote investment processes and, ultimately, to establish sustained prerequisites for rapid growth and improved economic efficacy.

The availability of scientific and technological achievements, research and innovation potential that can lead to progressive changes in technology and production, energy and material use are an important precondition of resource-saving advancement in the Republic of Belarus. National, sectoral and regional scientific and technological programs and projects lay the scientific foundation for addressing these issues.

Sustainable economic growth of the country will be ensured and maintained through modernization and restructuring of the real sector, technological and organizational renewal of industries to adapt them to internal and external specifics of the market economy. Sustainable economic growth of the national industrial complex should be based on high-technology, competitive and environmentally compatible production on conformity with world standards through active radical transformation of the existing scientific, technological and industrial potential.

The strategic goal of the investment and structural policy of Belarus is to attract investment, channel it into socio-economic priority areas and address ecological issues, as well as to develop a capital export policy through the use of competitive advantages of other countries by creating multinational companies, financial and industrial groups and other horizontal (cluster) structures. In order to facilitate investment flow into economy, further liberalization of foreign economic activities is necessary as well as the creation of mechanisms to insure investments against economic and political risks. It is necessary to ensure that public investment contributions further increase in size.

A transition of the Republic of Belarus to sustainable development in large measure depends on the active involvement of country regions in this process. In discharging their specific functions, local and regional authorities should contribute in their own way to sustainable development and improvement of the life quality of their residents. Thus, local administration plays a pivotal role in synthesizing all opinions and translating them into specific targets and programs, including into local sustainable development strategies (Local Agendas 21). In order to achieve the objectives and targets of regional sustainable socio-economic development and implement regional programs, it is necessary to provide for a set of interlinked measures and activities, economic mechanisms that would help address key issues, ensure concerted action of all stakeholders (including business and public organizations) involved in the implementation of each specific program.

The progress of Belarus toward sustainable development will decisively depend on the country's ability to expand contacts with the outside world. This is because the country's core industries and services sector are oriented toward the external market and there is substantial dependency of Belarus on imported raw materials. Deeper integration of Belarus into the world community presents a strategic goal. Foreign trade policy builds upon international rules and democratic standards, observance of international agreements, multi-directionalism and liberalization of foreign trade relations.

The primary objective of the Belarusian foreign economic policy is to effectively participate in the international labor division using competitive advantages of the country to raise the quality of life of its population through application of technological advances, dynamic development of economy and social sector while preserving the reproduction capacity of the country's natural complex as part of the Earth's biosphere in the interest of the present and future generations. For the purpose of implementing its foreign trade policy while being part of the global drive toward sustainable development, Belarus is envisaged to strengthen rapport with all players of international social, ecological and economic cooperation: participation in UN human development programs (UNDP), United Nations Environmental Programme (UNEP) and cooperation with the Global Environment Facility (GEF).

NSDS-2020 primarily focuses on the peculiarities and patterns of the forecast period, follow-up to the ?Agenda 21?, harmonization of social, economic and ecological development as equipollent and complementary elements in the ?man-environment-economy? balanced system.

The key sources of Belarus sustainable development should be: human capacity, scientific, production and innovation potential, natural resources, advantageous geographic location of the country, while the ?high intellect-innovation-prosperity? constitute top priorities. Transition to innovative development, system reforms in economy and society are the essential tasks of sustainable development in Belarus.

**Green Chemistry by Nano-Catalysis**

Polshettiwar, V., U. S. Environmental Protection Agency
Varma, R. S., U. S. Environmental Protection Agency

We envisioned a novel nano-catalyst system, which can bridge the homogenous and heterogeneous system, and simultaneously be cheaper, easily accessible (sustainable) and possibly does not require elaborate work-up. Because of its nano-size, i.e. high surface area, the contact between reactants and catalyst increases dramatically and essentially operate in the same manner as homogeneous catalysts. Concurrently,, due to their insolubility in the reaction solvent and paramagnetic nature, they can be separated out easily from the reaction mixture using an external magnet. Thus, these nano-materials have combined the advantage of both the system, and offered unique activity with high selectivity which will be highlighted.

**Achieving Comprehensive Social Impact Assessment**

Laurin, L., EarthShift
Hamilton, M. M., EarthShift

Just as environmental impact assessments miss important environmental affects and are often augmented by a risk assessment, the inclusion of social impacts in an assessment method will miss important social impacts. Because social impact assessment is a relatively new concept, there will be a learning curve over which the methods will get better and more comprehensive. In addition, social impacts are less tangible than environmental impacts, and may never be as well modeled using a life cycle assessment approach. These impacts can also be assessed using a risk assessment, this time an assessment of social risk.

Life Cycle Assessment (LCA) is often criticized for missing major environmental impacts. As an example, brominated flame retardants, even those which are no longer manufactured due to the possibility of risk to human health, are ignored by all major impact assessment methods. The risks to fish migrations in hydroelectric power generation is also ignored. To ensure that these important environmental impacts are not ignored in major decision-making, most LCA practitioners recommend a complimentary risk assessment to capture these other impacts.

Social LCA has even more potential for missing impacts. Cultural biases mean that one society's benefit is another's bane. We have seen, however, that social risk assessment as applied using Total Cost Assessment (TCA) has done an excellent job at digging out the potential social impacts of a behavior or decision. Applying TCA to several projects, social concerns have come to the forefront: concerns over noise, immediate human health impacts, disease transmittal from one location to another, and local employment, for example. While TCA is only one example of a social risk assessment, we will show how this method has brought the issue of social sustainability side by side with economic and environmental sustainability. By including this type of assessment as a compliment to the LCA methodology at the outset, we simplify the initial task for Social Impact Assessment Methods developers and give them a greater chance of success.

**Fuzzy Approachs to the Assessment of Sustainability Indices**

Imanov, G., Institute of Cybernetics of the National Academy of Azerbaijan Republic

This paper considers the sustainability of development level and environmental sustainability indices in the States. For the assessment of these indices we propose application of fuzzy logic inference method. Problem solution is demonstrated on the basis of International and Azerbaijani statistical data.

**Total Life-Cycle Approach to Sustainable Supply Chains**

Badurdeen, F., University of Kentucky

Abstract: Recent trends in globalization and outsourcing activities have created an increasing need for more efficient, effective and responsible business practices that focus on the triple bottom line (TBL). Conventional supply chains (SC) models are oriented towards increasing financial gains and return of investment (ROI). However, to promote sustainable supply chains (SSC) there is a need to focus on the TBL of economic prosperity, environmental protection and societal well-being. This requires an integrated and systematic approach that focuses on all life-cycle stages (pre-manufacturing, manufacturing, use and post-use) of the product. Such supply chain models require the application of innovative approaches like the 6R's (Reduce, Reuse, Recycle, Recover, Redesign and Remanufacture) to achieve closed-loop flow. Sustainability in the SC also depends on the agility and resilience of SC partners to proactively respond to unforeseen circumstances within the organization, in the industry or external sources. Thus, in the context of SSC's, risk management, too, plays a critical role. In this poster, we will present the concepts of SSC's (from the total life-cycle approach and the 6R's for sustainable manufacturing), compare and contrast existing SC frameworks, address their limitations, provide a framework for SC risk management followed by directions for future research.

**Linking Between ISO22628 and ISO/TS16949 to Building a Sustainable Innovation Excellence in Automotive Engineering**

Yusof, S. M., Universiti Teknologi Malaysia
Mat Saman, M. Z., Universiti Teknologi Malaysia
Zakuan, N., Universiti Tun Hussein Onn Malaysia

In the early 1990's the European Union identified End-of-Life Vehicles as a priority waste stream and the Directive was introduced to protect the environment. Automotive companies are being pushed by environmental awareness and legislation to recycle, remanufacture and reuse components at the end-of-life and also to reduce the quantity of manufacturing waste generated. Within this scenario, a design framework is needed for vehicle design processes in order to develop a sustainable concept. Besides that, the role of design in modern manufacturing is becoming even more important with companies adopting design framework as a profit generating business elements and also for measuring manufacturing performance. Hence, this paper discusses about two relevant standards in this case; ISO22628 is a standard that provides a method for determination of recyclability and recoverability for any vehicle design and ISO/TS16949 is a technical specification that represents a comprehensive quality management system for the global automotive industry to achieve world class levels of product quality, productivity, competitiveness and continual improvement. Based on these standards, the framework for sustainable innovation excellence in vehicle design and development processes was proposed in order to fulfill the current needs.

Keywords: automotive production standard, end-of-life vehicles directive, road vehicles standard, recyclability and recoverability, sustainable development

# Extraction of Bromide From Seawater and Bromination of Phenol with Seawater by Bio-Mimicking Catalysis

**Authors:**

Bharadwaj, S. K., Indian Institution of Technology Guwahati
Chaudhuri, M. K., Tezpur University

It is now well established that the naturally occurring bromoorganic compounds in marine aquatics are the result of Vanadium BromoPeroxidase(VBrPO) activity. The VBrPO enzyme catalyzes the oxidation of bromide by hydrogen peroxide followed by bromination of the organic compounds [1]. Although the reaction mechanism of VBrPO is not fully understood, it has been proposed that an active ?bromination species' is formed in situ, which could be ?Br+', ?OBr -' or ? Br3- '. The later has been isolated serendipitously in our laboratory which has led us to develop an eco-friendly and general route to the synthesis of quaternary ammonium tribromides (QATBs) [2].Taking cues from the bromoperoxidase activity and having successful synthetic procedure of QATBs in hand, we have been able to isolate quantitatively the solid tribromide by oxidizing the MgBr2 from seawater in presence of variety of catalysts. Herein, we are reporting a novel compound, viz. DmpzH[VO(O2)2dmpz] as one of the best catalysts and the experimental procedure for the extraction of bromide from seawater has been demonstrated [3]. A Variety of tribromides like TBATB, BTMATB, CTMATB, TEATB and TMATB has been synthesized from seawater quantitatively. Further, in view of wider use of bromo-organic compounds in the synthesis of a large number of natural products as well as in the manufacture of pharmaceuticals, agrochemicals and numerous industrially valuable products, bromination of organic compounds has received significant interest in recent years. For over a century, molecular bromine has been found most commonly and widely used brominating reagent as it is inexpensive and easily available. However, this classical bromination method suffers from an inherent disadvantage of 50% atom economy along with the production of one equivalent of hazardous HBr [4].Having encouraging result with the extraction from bromide from seawater and being known that tribromide might as well be an alternative for the bromination reaction, we have achieved direct bromination of phenol with sea water without adding external bromide, where tribromide has been generated in situ. Generally, bromination of phenol leads to mixture of bromophenol products (o-, p-, di-, tri- bromophenol). However, we have been able to achieve p-bromophenol selectively with seawater after conducting several reactions in presence of the aforesaid catalyst. 1. D.C. Crans, J. J. Smee, E. Gaidamauskas, L. Yang, Chem. Rev., 2004, 849. 2. M.K.Chaudhuri, U.Bora, S.K.Dehury, M.L.Kantam, B.M.Chaudhuri, US Patent 7005548, Feb. 28, 2006. 3. S.K.Bharadwaj, M.K.Chaudhuri, Indian patent( applied for). 4. P. B. De laMare, in Electrophilic Halogenation, Cambridge University Press, Cambridge, 1976, Chapter 5.

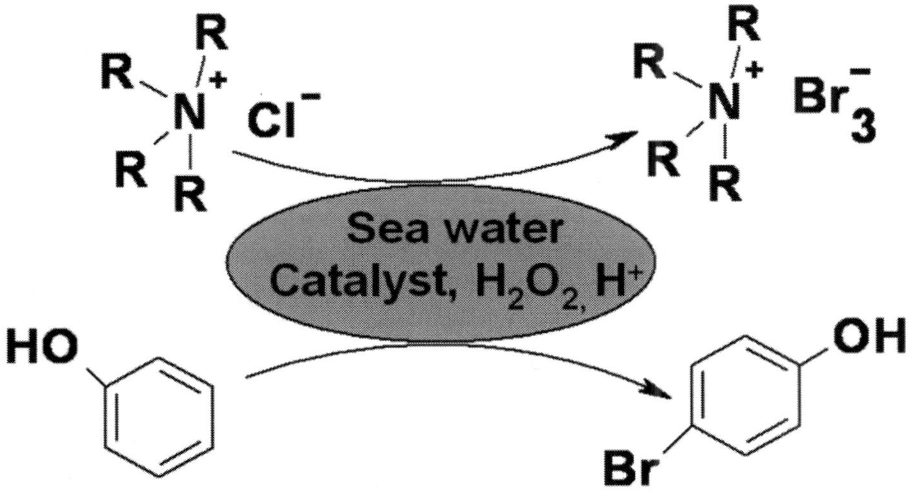


**Figure. Schematic representation of extraction of bromide from seawater as well as direct bromination of phenol with seawater**

-->

**Laboratory Experiments to Optimize a Packed Bed Reactor for the Production of Chemical Products From Sustainable, Waste Crude Glycerol**

Thomas, J., University of Kentucky
Hansrote, S., University of Kentucky
Richardson, L., University of Kentucky
Seay, J. R., University of Kentucky

Student Poster Session Presentation

The process for the production of biodiesel from vegetable oils yields large quantity of crude waste glycerol as a by-product. The glycerol by-product has the effect of reducing the overall carbon utilization of the biodiesel process. However, if the crude glycerol produced could be used to create an industrially important yet sustainable product the carbon utilization as well as the overall process economics of the biodiesel could be improved.

It is well known that a range of chemical products can be produced from crude glycerol. Based on established literature data, a bench scale reactor was designed and built to produce an array of C3 compounds from waste crude glycerol. The reactor has the ability to produce a variety of C3 compounds, such as acrolein, 1,2-propanediol, 1,3-propanediol, hydroxyacetone, and acrylic acid. The purpose of this research is to conduct a series of experiments to determine the optimum operating conditions for this reactor. Optimization parameters include: flowrate, composition, temperature and catalyst concentration. Additional experiments to develop a procedure to regenerate the catalyst in place are also included.

The laboratory results generated from this project will be used to provide input for a computer simulation of an industrial scale glycerol dehydration process including the waste and recycle streams and all separation processes necessary to yield a marketable product. Using the computer simulation, a cost effective set of parameters will be determined and the results will be tested using the laboratory small scale reactor for comparison.

**Developing a Flexible Economic Model for the Production of Chemical Products From Waste Crude Glycerol**

Gish, J., University of Kentucky
Ramey, K., University of Kentucky
Seay, J. R., University of Kentucky

Student Poster Presentation

Biodiesel is a sustainable and clean source of energy both in the manufacturing process and upon consumption. Additionally, it is produced from renewable resources such as vegetable oils or animal fats. However, the process of manufacturing biodiesel from vegetable oils is the generates glycerol as a byproduct. In the biodiesel production process, 1 kg of glycerol is generated for every 9 kg of biodiesel produced.

Previous research has identified numerous industrially important chemical products that can be manufactured from glycerol. In particular, many C3 compounds, currently manufactured from propylene can also be manufactured from bio-based glycerol. Some of these products include: 1,2-propanediol, 1,3-propanediol, acrolein, hydroxyacetone and acrylic acid. Developing process models for the production of these, and other products, from glycerol will be useful because it will identify which products are the most economically viable and establish cost targets for switching production from crude oil derived propylene to glycerol.

This contribution is focused on development the process simulation models that will serve as the basis for a flexible economic model for determining the optimal product slate for a multipurpose reactor based on glycerol dehydration chemistry. The process simulation models produced from this research will also be based on minimizing the environmental impact while providing industries with economically efficient and technically feasible processes. To achieve this, the Waste Reduction (WAR) algorithm, developed by the U.S. Environmental Protection Agency, will be applied to evaluate the potential environmental impact (PEI) of chemical process developed as part of this research. This analysis will be part of a holistic approach performed by applying process simulation in combination with laboratory experiments for the conceptual design of chemical processes.

**Benchmarking Sustainability for Lignocellulosic Conversion**

Inman, D., National Renewable Energy Laboratory
Aden, A., National Renewable Energy Laboratory
Davis, R., National Renewable Energy Laboratory
Chum, H., National Renewable Energy Laboratory
Hsu, D., National Renewable Energy Laboratory
Heath, G., National Renewable Energy Laboratory
Mann, M. K., National Renewable Energy Laboratory
Foust, T., National Renewable Energy Laboratory

Biofuels derived from both starch and lignocellulosic feedstocks are expected to contribute 36 billion gallons per year by 2022 according to the Energy Independence and Security Act (EISA) of 2007. Questions regarding the overall sustainability of such an unprecedented increase in bioethanol and biofuels production have focused, to a large extent on the feedstock production side of the bioethanol supply chain because of issues of direct and indirect land use changes. Tools for their assessments are still evolving. Other issues include food versus fuels, hunger and poverty issues, among others, which make it imperative to explore lignocellulosic biomass. Recent studies have illustrated the importance of the biorefinery to the overall life cycle greenhouse gas emissions and fossil energy demand. It is imperative for each component of the ethanol supply chain to be optimized with regard to key metrics of sustainability as well as the overall system. The National Renewable Energy Laboratory is currently using process engineering and life cycle assessment coupled to analyses of other parameters to establish benchmarks for sustainability within the biorefinery. These metrics will be used to spur novel process and system improvements and to ultimately establish sustainability targets. These targets will complement traditional technoeconomic analyses that have so far guided selection of key technical barriers for cost optimization. Key areas under investigation include greenhouse gas (GHG) emissions (includes energy sources and energy efficiency), water consumption, and refinery waste treatment needed in both the biochemical conversion of corn stover to ethanol and the thermochemical gasification of mixed woods to ethanol and mixed alcohols. Examples of sustainability benchmarks for a few biorefinery processes and potential improvements will be presented.

*This work is funded entirely by the United States Department of Energy's Office of Energy Efficiency and Renewable Energy's Biomass Program.

**Using the Box-Benkhen Design to Statistically Model Hydrogen Production During Glucose Fermentation in the Presence of Oleic Acid**

Ray, S., University of Windsor
Lalman, J. A., University of Windsor

Hydrogen ($H_2$) has been projected as a sustainable fossil fuel replacement and its production from non-conventional routes has gain a significant amount of research attention over the past several years. Over the past several years, many governments have emphasized the need for prioritizing the development of biological processes such as biophotolysis and anaerobic fermentation to produce sustainable energy source. These biological methods are able to produce hydrogen by enzymatic degradation of organic waste and agriculture residues [1]. Among the various biological routes of hydrogen production from organic compounds, the light-independent dark fermentation is considered economically more favorable due to its substrate non-specificity and faster reaction rates [2]. Moreover, recent studies have shown that the anaerobic process is more efficient than many biological processes in terms of net energy utilization [3].

In anaerobic dark fermentation complex organic substrates (such as hexose) are fermented to hydrogen and organic acids. However, the process bottleneck is attempting to achieve the theoretical hydrogen yield (4 mol $H_2$/mol hexose) because of hydrogen consumption by hydrogen consuming microorganism [4]. Hydrogenotrophic methanogens (MB), sulfate-reducing bacteria (HSRB) and homoacetogens are the principal hydrogen consumers in anaerobic mixed microbial systems. The presence of hydrogen consumers causes a loss in electron equivalence by diverting electrons to cell biomass formation and facilitating incomplete oxidation of reduced by-products [5]. Due to lower free energy of hydrogen utilization for HSRB (-38.1 kJ/mol) compared to that of methanogens (-32.7 kJ/mol), HSRB are more efficient hydrogen consumers than methanogens. However, HSRB also need at least a $C_2$-organic carbon source in addition to hydrogen for cell synthesis and they are often limited beyond a threshold substrate level [6]. Compared to homoacetogensis (-5 to -6 kJ/mol $H_2$), methanogenesis (-9 to -12 kJ/mol $H_2$) is energetically a more favorable route for hydrogen consumption. In a competitive environment of hydrogen utilization, methanogens are able to outcompete homoacetogens [7]. Hence, increasing the hydrogen yield in mixed anaerobic fermentation by controlling the hydrogen-consuming methanogenic populations is of significant interest to many researchers

Methanogenic growth control in a mixed consortium can be effected either by modulating the operational conditions of the bioreactors or through addition of chemical growth inhibitors or both in combination. Koster *et. al.,* reported inhibition of methanogenic population in a mixed consortium using long chain fatty acids (LCFAs) [8]. Lalman *et. al.,* observed that the addition of LCFAs, such as linoleic acid (LA) (C18:2) and oleic acid (OA) (C18:1) resulted in the accumulation of acetic acid [9,10]. The inhibitory effect of different linoleic acid (LA) (C18:2) levels was examined by Chowdhury *et. al.* (2007) and these researchers reported that increasing the LA concentration decreases the methanogenic activity and diverted a large fraction of electrons towards hydrogen production [11]. Ray *et. al.* (2008) also reported that hydrogen production from a mixed mesophilic culture was favoured when LA was added under low initial pH conditions [12]. The study also reported improved hydrogen production on successive glucose injections. Evidently the choice of LCFA and favourable level of operating variables is of utmost importance for the enhancement of hydrogen yield in batch cultures. To date, no study which has assessed the effect of experimental variables and OA concentration on the hydrogen yield of mixed cultures.

Experimentally, the optimum levels of experimental settings can be located by a single factor optimization method. However, the maximization of hydrogen yield through a single factor optimization is often considered less advantageous than optimization using statistical experimental design. Among the available statistical designs, a fractional factorial design (FFD) is a common choice for investigating hydrogen yield. However, FFD is neither very accurate (lacks rotatability) nor economical (require large number of experiments) [13]. Alternately, response surface optimization using the Box-Benkhen design (BBD) has been seldom considered in optimization studies for fermentative hydrogen production [14]. Hence the objective of the present study is to maximize hydrogen production in a mixed mesophilic culture at 37°C by assessing the effects of initial pH, OA concentration and number of glucose injection using a three factor three level BBD.

The experimental levels for each factor (Table 1) were selected based on literature values and the results from preliminary experiments. The hydrogen-yield data from BBD experiments conducted in batch using mixed cultures were analyzed using Minitab 15 statistical software (Minitab Inc. State College, PA). Analysis of variance (ANOVA) and regression analysis was performed to derive a response surface model. Three dimensional surface plots were used to investigate the factor effect on design surface and locate the optimum factors conditions for maximum hydrogen yield.

**Table 1:** Factors and levels selected for the experimental study

| | **Factors** | **Levels** | | |
|---|---|---|---|---|
| | | **1 (Low)** | **2 (Middle)** | **3 (High)** |
| 1 | **LA concentration ($g\times l^{-1}$)** | 0 | 1 | 2 |
| 2 | **Initial pH** | 5.0 | 6.0 | 7.6 |
| 3 | **Number of glucose injections** | 0 | 1 | 2 |

Anaerobic cultures utilized for this work were sourced from an ethanol manufacturing facility (Chatham, ON) and from a municipal wastewater treatment facility (Chatham, ON). The two inoculum were mixed in a 70:30 ratio (ethanol: municipal) in a semi-continuous reactor and maintained at 37±1°C. The 4-l semi-continuous reactor (denoted as reactors A) with 2,000 $mg\times l^{-1}$ VSS were prepared by diluting the mixed inoculum. This reactor served as source of anaerobic mixed culture for all experiments. A basal medium, adapted from Weigant *et. al.* was used for feed preparation and diluting the inoculum [15]. The experimental methods reported in this work were adapted from earlier studies published in the literature (9, 10, 16). Each bottle was filled with 50 ml of basal media plus culture (2,000 mg mg.$l^{-1}$ VSS) in an anaerobic glove box (COY Laboratory Products Inc., MI) under a 70% $N_2$/30% $CO_2$ (Praxair Inc. ON) atmosphere. A 50,000 mg.$l^{-1}$ OA stock solution was prepared using OA melted *au bain-marie* in hot NaOH and 100,000 mg×$l^{-1}$ glucose stock solution was prepared by dissolving glucose in MQ water. The pH was adjusted with 1M HCl or 1M NaOH. All controls and culture for the experiment were prepared in triplicates in 160 ml serum bottles. The bottles were sealed in the anaerobic glove box with Teflon[â]-lined silicone rubber septa and aluminum crimp caps. Each bottle was over-pressurizing with 20 ml of 70% $N_2$/30% $CO_2$ to avoid the formation of a negative pressure in the headspace during sampling. The bottles were agitated using an orbital shaker (Lab Line Instruments Model 3520, IA) at 200 rpm and maintained at 37±2°C over the duration of the study. Headspace gas samples (25ml) were withdrawn at regular intervals and analyzed using a Varian-3600 (Varian, Palo Alto, CA) gas chromatograph (GC) configured with a thermal conductivity detector (TCD) to measure the quantity of hydrogen produced. A 2-m long x 2-mm I.D. Carbon Shin column (Alltech, Deerfield, IL) was used for conducting the analysis. The GC injector, detector and oven temperatures were set at 100°C, 200°C and 200°C, respectively. The nitrogen carrier gas flow rate was set at 10 ml $min^{-1}$. The detection limit for hydrogen was 0.0032 kPa.

Increasing OA concentration and low initial pH conditions were observed to enhance the hydrogen yield. Higher hydrogen yields were recorded with two consecutive glucose injections as reported in the literature for cultures receiving LA. Based on the Box-Benkhen technique, a response surface model was developed for computing the hydrogen yield (mol $H_2$×mol $glucose^{-1}$) from glucose fermentation. The hydrogen yield in the response surface model was expressed in terms of the OA concentration, initial pH and number of glucose injections (three design variables). Significant interactions were observed to exist between the experimental factors. The model was observed to have a 92.5% correlation (R-square of 85.5%) over the entire factor space under evaluation. A maximum hydrogen yield of 2.61 mol $H_2$×mol $glucose^{-1}$ was predicted with 2 g×$l^{-1}$ OA at an initial pH of 5.0 for two glucose injections. The observed hydrogen yield in experiments conducted near the computed optimum factor setting was 2.04 ± 0.27 mol $H_2$×mol $glucose^{-1}$.

**References:**


1. Das *et al.* (2001). *Int. J. Hyd. Energy.* 26 (1), 13-28.

2. Kargi *et al.* (2006). *Enzym. Microb. Technol.* 38 (5) 569-582.

3. Hallenbeck *et al.* (2002). *Int. J. Hyd. Energy.,* 27 (11-12), 1185-1193.

4. Manish *et. al.* (2008). *Int. J. Hydrogen Energy.* 33 (1), 279 - 286.

5. Oh *et. al.* (2005). *Water Res.* 39 (19), 4673-4682.

6. Liamleam *et. al.* (2007). *Biotechnol. Adv.* 25(5), 452-463.

7. Conrad *et. al.* (1990). *Arch. Microbiol.* 155 (1) 94-98.

8. Koster *et. al.* (1987). *Appl. Env. Micbiol.*, 53(2), 403-409.

9. Lalman *et. al.*(2000). *Water Res.* 34 (17), 4220-4228.

10. Lalman *et. al.* (2001). *Water Res.* 35 (12), 2975-2983.

11. Chowdhury *et. al.* (2007). *J. Environ. Eng.* 133 (12) 1145 ? 1152.

12. Ray *et. al.* (2008). *J. Environ. Eng.* 134 (2), 110-117.

13. Myer *et. al.* (2002). *Response surface methodology: Process and product optimization using designed experiment*, second ed., John Wiley and Sons, New York, 343-350.

14. Wang *et. al.* (In Press). *Int. J. Hydrogen Energy*. doi:10.1016/j.ijhydene.2008.10.008.

15. Weigant *et. al.* (1985). *Biotechnol. Bioeng.* 27 (11), 1603-1607.

16. Lalman *et. al.* (2002). *Water Res.* 36 (13), 3307-3313.

**Effects of Alternative Mixed Alcohol Synthesis Configurations On Ethanol Production by Indirectly-Heated Gasification of Lignocellulosic Biomass**

Zhu, Y., Pacific Northwest National Laboratory
Gerber, M. A., Pacific Northwest National Laboratory
Jones, S. B., Pacific Northwest National Laboratory
Stevens, D. J., Pacific Northwest National Laboratory

As a sustainable energy source, the conversion of biomass to alternative liquid fuels plays an important role in solving oil shortage and environmental problems being confronted today. Currently, these applications are mostly met through the conversion of grain products into fuel-grade ethanol that can be blended with gasoline. However, while these forms of biomass can supply significant quantities of transportation fuels, other more widely available biomass forms, specifically lignocellulosic biomass forms, also need to be used to have a major impact on domestic supplies of transportation fuel. At this time, lignocellulosic biomass gasification technologies have been developed to the point of large-scale demonstrations. However, widespread commercialization of large-scale biomass gasifiers or their integration with ethanol synthesis technologies has not been realized because of technical and economic challenges. With these technologies still in early development phase, system modeling and economic analysis should be implemented to provide insights for decision making about the research and development of these technologies. Techno-economic assessments have been conducted for ethanol production from biomass systems. Limited sensitivity analyses were performed to evaluate the impacts of alternative configuration assumptions on overall system performance and costs. The purpose of this study is to investigate the technical and economic sensitivities of biomass gasification based ethanol production systems related to alternative mixed alcohol synthesis configurations, and thus to identify potential opportunities for the system improvement and optimization.

In this study, a systematic analysis method is used to investigate the impacts of alternative mixed alcohol synthesis configurations on the ethanol production from indirectly-heated gasification of biomass. The system consists of five major processes: (1) gasification, (2) syngas purification, (3) syngas adjustment, (4) mixed alcohol synthesis and product purification, (5) steam cycle. In this system, wood chips are converted to syngas in the gasifier. The syngas is cleaned and the $H_2/CO$ ratio is adjusted. The syngas is then converted to mixed alcohols and separated into products. Unconverted syngas are used as fuels for processes and electricity production. Alternative mixed alcohol synthesis configurations with difference process assumptions are examined in this study. The effects of different methanol recycle ratios, different syngas recycle ratios, and alternative recycle syngas injection locations on the system performance and costs are evaluated. For each configuration, a process model is developed in CHEMCAD to estimate heat and material balances. Cost analyses were implemented based on inputs from the process models.

The simulation results are analyzed and discussed. The major performance analysis results include carbon conversion efficiencies, power consumption and output, and final product yields. The cost analysis results include estimated capital and operating costs, and ethanol selling prices. Results analysis indicated that with the methanol recycle increasing from 0% to 90%, the ethanol selling price decreases approximately 14% resulting from the decrease of the ethanol yield. The impacts of syngas recycle ratios and recycle syngas injection locations are also discussed. By comparing the performance and cost results of alternative configurations, potential opportunities for improving the performance and costs of gasification based biomass-to-ethanol systems are identified. By identifying those changes that have the greatest impacts, insights for decision making on research and development priorities of these technologies are provided. Research can be more effectively directed toward achieving economic viability by focusing research in the key areas and setting processing milestones to measure technical progress.

In this paper, the process design basis and modeling approach are described. Major performance and cost results of case studies are listed and analyzed. The results of alternative configurations are compared and the impacts of different process assumptions are discussed.

Keywords: Biomass, Gasification, Syngas, Mixed Alcohol

**Sustainable Infectious Disease Surveillance and Biosafety**

Hathaway, K., Armed Forces Research Institute of Medical Sciences

Maintaining human, animal, and environmental health is of critical importance to sustainable development around the globe. Establishing, enhancing, and maintaining infectious disease surveillance labs to protect people and animals from disease requires time and in addition to monetary capital for construction, quality biosafety laboratories require significant investment in personnel, training, and ongoing maintenance. Current efforts to develop and enhance laboratory capacity around the world have improved global biosafety, yet challenges remain.

Biosafety laboratory regulations will be reviewed along with past projects to develop individual labs and regional capacity. Construction of new state of the art facilities will be reviewed along with the training and support staff required to maintain some of the world's most established biosafety laboratories. The cost of both construction and operation of a biosafety lab will be reviewed, along with the potential impact of these costs in resource limited settings. Options to improve sustainable laboratory construction (materials and construction processes), operation (balancing energy efficiency and biocontainment) and future development (planning for renovation and arranging to analyze high-risk samples in global collaborating centers) will also be put forth for discussion with the biosafety and sustainable development community.

**Transforming Ourselves to Live in Compression**

Hall, R. W., Association for Manufacturing Excellence

We live in globally perilous times, for the 21st century is a turning point. Physical economic expansion began 500 years ago with European colonization. It cannot continue. From a global perspective, the earth has finite resources, but our business and economic legacies assume that more resources are always available ? somewhere. These legacies evolved to promote expansion. They still grip us, so transforming ourselves to limit consumption while simultaneously increasing human quality of life for everyone is the supreme human challenge of Compression.

Entrepreneurs seek a magic technical fix to let us preserve our profligate past, but no single countermeasure is likely to suffice because the old system has myriad challenges, in five categories: 1. Limited natural resources. 2. Over-consumption: The system's incentives promote using more and more. 3. A precarious environment: no one can keep up with all the problems in detail; global warming is merely the one most publicized. 4. Pushback: people whose way of life or sense of fairness are injured in the mad rush have resisted the expansionary system for the past 500 years, and those issues have not gone away.

But the key challenge is number 5: transforming ourselves, our systems, and our thinking to deal with all these problems rolled together; here labeled Compression. Taken together, all these challenges make it unlikely that any single technological fix is going to resolve them. We need to devise a different approach, one that recognizes that nature is not a backdrop to the human economy. Our economic activity is a subset of nature, affecting it and affected by it.

Recognizing this changes everything: The purpose of work is to assure human survival with an acceptable quality of life for all ? quality over quantity. It upsets existing economic systems, capitalist and socialist, for both assume unending physical expansion. Arbitrarily quantifying this challenge is a first step toward determining how to take practical action ? things we can do:

Improve industrial society quality of life; extend it to everyone globally while using less than half the energy and less than half the virgin raw materials of today, while reducing toxic releases nearly to zero.

All of our problems are ameliorated by simply consuming less ? even if we were able to tap huge sources of non-fossil, non-solar, clean energy. And cutting global consumption in half implies a much stiffer challenge than that in high-consumption parts of the world. This sounds like doomsday to those living in the comfort of the existing order with no idea how to go about making necessary changes. That's why changing ourselves to learn to change technologies and business models faster than ever before is our core challenge.

Rapid learning requires a work organization (here called a vigorous learning enterprise) in which most people are professionally dedicated to its social mission; and financial solvency, while necessary, is secondary. It's main purpose is peak performance addressing all work critical to human survival and quality of life. Governance must be on behalf of all stakeholders, not just owners, blurring distinctions between public and private funding.

This is not a pipe dream. Real people have actually performed almost in this way. Ideas for it are blended from the best operational organizations today: servant leadership, professional attitudes, lean and quality concepts applied by all, technological proficiency, and flexibility to do what is needed with minimal command and control. A vigorous learning enterprise is really a communication network sized to maximize human learning speed, concentrating on a social mission, not profit maximization.

However, our biggest obstacle is our old legacies that evoke tribal instincts and provoke dysfunctional competition instead of speeding our learning. Many of these are so normal to us that we seldom think about them. Much of the presentation will delve into these ?hidden legacies? in our concepts of work and how to manage it, on the thought that if we are to make big changes, first we have to recognize what needs changing. That is, learning to emphasize quality over quantity is not natural. It's a step up in civil self-discipline. It can be done if we have the will.

**Modeling of Adsorption Data On the Removal of Heavy Metals From Industrial Waste Water**

Nivas, A., BANNARI AMMAN INSTITUTE OF TECHNOLOGY

Heavy metals are stable and persistent environmental contaminants. They cannot be degraded or destroyed. The main anthropogenic sources of heavy metals are industries. The removal efficiency of heavy metals by preparing activated carbon from sugar mill sludge is determined. The carbon has been used as adsorbent .The experiment was carried out manually by batch process and column flow system. The results obtained are processed statically by developing a program to run in MATLAB software and the removal efficiency obtained through MATLAB processing was more than 90%.

This experimental study has been carried out to develop an economical method of heavy metal removal, so that even small industries can adopt this method for their waste water treatment and hence they can prevent the polluted water entering the stream.

The results are provided in the requisite tabular as well as graphical forms.

**Modeling the Value Chain Sustainability of Forest Resource-Based Products**

Halog, A., University of Maine

A modern sustainable economy has highly interconnected industries which optimally use both renewable and non-renewable resources for sustainable production and consumption of goods and services. Industrial ecology-based development can lead to a sustainable co-existence of economy, technology and natural environment. However, this requires transformative, integrative, life cycle oriented framework and interdisciplinary approaches to shift our human actions and behaviors towards sustaining our current and future generations development. Service and manufacturing industries are evolving to become environmentally responsible, bio-based, and capitalizing information and communication technologies (ICTs) to develop industrial ecology-modeled supply (value) chains. Information technology has negatively affected the forest based industries such as lesser consumption of newsprint and office papers. This negative impact can also stimulate the pulp and paper industry to refurbish its operations and produce highly value added products. US forest industry can find ways to combine wood resources with ICT in new innovative ways such as supporting the production of bioproducts and biofuels. These newly found business opportunities can intensify the ongoing restructuring of the forest sector and make this industry viable in the face of global challenges. ICT is considered significant enabler in the creation of value chain networks. The goal of this project is to develop an ICT enabled framework and conceptual design model for creating sustainable supply chains of forest-based bio-products. ICTs such as electronic data interchange (EDI) and ?cyberinfrastructure? can help in increasing the productivity and viability of an industrial sector. EDI-based model is also applicable to the current challenging pursuit of broadening and deepening of life cycle assessment (LCA) method to account economic, social and other dimensions of sustainability. EDI-based system can support secure exchange of relevant information of trading companies within a supply chain and thus enables comprehensive and reliable life cycle sustainability assessment of products and services. Through this novel and robust approach, information regarding energy efficiency, economic, socio-economic and environmental performance, as well as other sustainability issues can be exchanged within the supply chain networks. Thus, this work can be the basis in developing a "cyberinfrastructure" enabling framework to derive novel scientific theories and knowledge.

**Sustainability Measures and the Global Reporting Initiative (GRI)**
Henderson, R., ABS Consulting

Sustainability requires an organization to pursue a strategy that encompasses social, environmental and business performance. Superior organizations also report their sustainability performance. Key components of a sustainability report are the measures used to determine progress towards becoming a sustainable organization.

A sustainability report should provide a balanced and accurate representation of the sustainability performance of the reporting organization, including both positive and negative contributions.

The Global Reporting Initiative (GRI) was established to promote sustainability performance reporting using a standardized format that promotes adequacy, consistency and comparability across reports. GRI's vision is that reporting on sustainability by all organizations become as routine and comparable as financial reporting. The GRI is considered the ?Gold Standard? for sustainability reporting.

The GRI network accomplished this vision by developing, continuously improving, and building capacity around the use of a Sustainability Reporting Framework, the core of which are the Sustainability Reporting Guidelines. Other components in the Reporting Framework are Sector Supplements and Protocols. This reporting guidance - in the form of principles and indicators - is provided as a free public good.

To ensure the highest degree of technical quality, credibility, and relevance, the GRI Reporting Framework was developed and continuously improved through intensive multi-stakeholder engagement that involved reporting organizations, sector experts, and information seekers, who together develop review, and test content for the Reporting Framework.

Sustainability Measures as defined by the GRI G3 Guidelines:

The GRI G3 reporting framework structures its indicators by category, aspect, and the importance of the indicator across and within sectors.

Categories

The GRI categories are the triple bottom line categories: economic, environmental, and social. The social grouping is sub-categorized into Labor Practices, Human Rights, Society, and Product Responsibility. A given category may have several Aspects, which in turn may have multiple indicators.

Aspects

These are the general types of information that are related to a specific Indicator category (e.g., energy use, child labor, customers).

Indicators

These are qualitative or quantitative information about results or outcomes associated with the organization that is comparable and demonstrates change over time. Examples of GRI measures include:

1) Economic:

a. EC1 Direct economic value generated and distributed, including revenues, operating costs, employee compensation, donations and other community investments, retained earnings, and payments to capital providers and governments.

b. EC2 Financial implications and other risks and opportunities for the organization's activities due to climate change.

2) Environmental:

a. EN1 Materials used by weight or volume.

b. EN2 Percentage of materials used that are recycled input materials.

c. EN4 Indirect energy consumption by primary source.

d. EN16 Total direct and indirect greenhouse gas emissions by weight.

3) Social:

a. LA1 Total workforce by employment type, employment contract, and region.

b. LA7 Rates of injury, occupational diseases, lost days, and absenteeism, and number of work related fatalities by region.

c. HR2 Percentage of significant suppliers and contractors that have undergone screening on human rights and actions taken.

d. SO2 Percentage and total number of business units analyzed for risks related to corruption.

e. PR1 Life cycle stages in which health and safety impacts of products and services are assessed for improvement, and percentage of significant products and services categories subject to such procedures.

Core and Additional Indicators

Core Indicators are considered to be basic, required measurements applicable to all organizations. Additional indicators are optional and may represent leading practice.

Indicator Protocol

The GRI indicator protocol is essentially a recipe for assembling the data and support materials for an indicator. Each indicator has a unique and detailed protocol, which has been developed based on practical reporting experience across and within sectors.

This paper/presentation will take audience through the process of selecting and using well established, and globally accepted, sustainability measures.

**Reduction of World's Fossil Fuel Consumption through Cargo Containers Superficial Photovoltaic Solar Energy Harvesting**

Rodríguez-Ruiz, Sr., C. G., CENIPRO

Fossil fuel consumption rate is unsustainable for the planet not only because of global warming, but because in approximately 41 years current consumption trend will surpass production rate. The objective of this paper was to analyze technical feasibility of exploiting world's multimodal cargo container fleet top surface via photovoltaic (PV) solar energy harvesting along the entire transportation supply chain, whenever meteorological and logistics conditions permit. Cargo transportation power consumption will allow little or no need for energy storage and its global dynamic could enable constant PV renovation for efficiency re-enhancing. World's PV's production was exponentially and logistically forecasted, and the results show that unless growth rate ceases to increase by 2012, PV's production will allow harvesting more than 65% of world's cargo container fleet by 2030 to provide with an only 15% solar cell efficiency, more than 1.78 TW of power per year, supplying 7.58% of world's energy consumption ?mainly from the most pollutant transportation fuels such as fuel-oil or diesel? and curbing over 3 billion tons CO2 emissions per year.

Keywords: Solar energy, sustainable transportation, fossil fuel consumption, renewable energy.

**Best Practices in Sustainable Energy From the Public and Private Sectors**

Marks, S., Brown and Caldwell
Krishnayya, M., Brown and Caldwell
Trueblood, D., Brown and Caldwell

Best practices in sustainable energy will be presented from both the public and private sectors to share practical ways to reduce energy usage in a cost effective and comprehensive manner.

The first item to be presented will be a guidebook that was prepared by Brown and Caldwell for Sonoma County, CA. The guidebook shares a conceptual process for developing, evaluating, and implementing best energy practices. It also shares proven best practices for use by Sonoma County agencies and local governments in Sonoma, Napa, Marin and Solano Counties that lack the resources and/or expertise to launch Utility Management Programs. The guidebook is being used by the Energy and Sustainability Division (ESD) of the General Services Department of the County of Sonoma to support a 24-month implementation of this innovative regional strategy. The Utility Management Project disseminates proven and tested processes, tools, techniques and policies to most efficiently and effectively manage internal energy, water and waste related utilities in ways that reduce energy utilized and reduce greenhouse gas emissions. Through replication of these best practices, this project will measurably reduce greenhouse gas (GHG) emissions in the region.

The Utility Management Project includes development and dissemination of the guidebook and will be followed by four half-day workshops to which nearly 200 agencies throughout the region have been invited and expert technical assistance provided to participating agencies for implementing these utility management tools. One government agency will be selected to participate in a pilot project to demonstrate and test the effective replication and implementation of these utility management tools.

The Sustainable Energy Practices Guidebook was created to help public agencies benchmark their greenhouse gas emissions and devise workable, achievable plans for reducing them. During the course of this project, we talked with staff in Sonoma County and technical experts across the country to uncover case studies and proven best practices that are relevant to the needs of government agencies and applicable to utility-based programs. The result is a structured set of sustainable practices and suggested policies, quantitative and financial analysis tools, and step-by-step ?how-to? guidance for establishing and implementing an effective greenhouse gas reduction program.

This part of the presentation will be followed by a description of the energy ?treasure hunt? process used by several Fortune 500 companies to identify, cost, and implement energy reduction programs. This best practice from the private sector has been a very effective process to comprehensively reduce and minimize energy savings. It can be applied to a single production process or applied as an overall program on a global basis.

This process typically lasts 3 to 5 days and involves establishment of a number of utility focused teams from the company that intensively review focused areas of production and support operations to identify sources of energy uses and waste. The process is overseen by a facilitator, who conducts a daily review of the results and also provides the teams with tools to characterize the energy usage and potential alternatives. The event starts during a time period when minimal production is occurring to identify baseline energy usage and potential waste.

At the end of the treasure hunt event, all of the teams' results are presented to senior managers. The managers are required to either approve, disapprove, or request more study for each potential energy saving project that is identified before the end of the outbrief. Typically, 5 to 10 major energy savings projects are selected for implementation and total cost savings are in the tens to hundreds of thousands of dollars ann

**Framing a Business Case for Sustainability**

Allevato, F., Det Norske Veritas
Stahl, C., Det Norske Veritas AS

Many companies fail to make the connection between healthy ecosystems and the attainment of their economic goals (World Business Council for Sustainable Development, 2008). This paper discusses proactive management of environmental resources, an often overlooked business opportunity, using the ecosystem services approach. Det Norske Veritas (DNV) has been working closely with the World Business Council for Sustainable Development (WBCSD) and the World Resource Institute (WRI) in shaping this approach so that environmental management practices can be optimized by better understanding the relationships between business goals and the ecosystem services they impact and depend on.

Understanding ones' ecosystem impacts and dependence, however significant, is only a first step towards achieving sustainable ecosystem use and management. It should be followed by efforts to internalize and account for the full value of ecosystems as well as the costs associated with ecosystem service loss. Only then will strong incentives be present for internalizing these findings and incorporating them into corporate strategy.

Despite the fact that ecosystem services underpin markets everywhere, ecosystem services are rarely traded and are often viewed as externalities. As a consequence, many ecosystems and the services they provide are undervalued. The use and degradation of ecosystems appear to carry either very low or zero costs, and hence there is a need to support sustainable businesses in valuing their ecosystem relationships and assets.

As a result, the apparent business case for progressive action may be weak because of the lack of tools and expertise to quantify the worth of ecosystems in terms of Net Present Value. These unquantified environmental risks and opportunities can be estimated however using a combination of operations knowledge, ecosystems knowledge, and a systematic business approach.

The business cost of failure to manage environmental resources is often hidden in losses due to current inefficiencies, unaccounted future expenditures, and risks and opportunities that arise from ecosystem degradation. The cost of an environmental aspect includes today's lost energy, resources, and opportunities, cumulative losses over time, and losses resulting from related environmental aspects. Therefore, the profitability and sustainability of many industries is inextricably linked to sustainable care of the ecosystem services they impact and depend on.

**The Sustainable Enterprise: Integrating Science and Business Models**

Genaidy, A., University of Cincinnati
Sequeira, R., University of Cincinnati
Tolaymat, T., U.S. EPA
Rinder, M., WorldTek
Curran, M. A., BAMAC Ltd.

Background ? Secondary lead smelters (SLS) represent an environmentally-challenging industry as they deal with toxic substances posing potential threats to both human and environmental health, consequently, they operate under strict government regulations. Such challenges have resulted in the significant reduction of SLS plants in the last three decades. In addition, the recycling of lead has been on a steep decline in the past ten years although the amount of lead recovered has remained virtually unchanged. Therefore, one may wonder whether sustainable development can be achieved among SLS.

Objectives and Specific Aims ? The primary objective of this study was to determine whether a roadmap for sustainable development can be established for SLS. The following aims were established in support of the study objective: (1) to document the challenges and opportunities for the U.S. secondary lead smelting industry; (2) to conduct a systematic review and a critical appraisal of models of sustainable systems with a particular emphasis on SLS; and (3) to determine the potential for a sustainable development roadmap for SLS.

Methods ? An evidence-based methodology was adopted to achieve the study objective. A comprehensive electronic search was conducted to implement the aforementioned specific aims. Inclusion criteria were established to filter the scientific papers and reports. The relevant articles were closely scrutinized and appraised to extract the required information and data for the possible development of sustainable development roadmap.

Results and Discussion ? The search process yielded a number of research articles which were utilized in the systematic review. Two types of models emerged: management/business and science/mathematical models. Although the management/business models explored actions to achieve sustainable growth in the industrial enterprise, science/mathematical models attempted to explain the sustainable behaviors and properties of simple economic-ecological systems. In few cases, management models examined the behaviors of sustainable systems.

An integrated model has yet to be developed to synthesize the common and unique perspectives of different researchers within the context of the sustainable enterprise with a clear vision and objectives. Also, the mathematical representation of the sustainable enterprise arising from science models has to be established. In this regard, the interaction of models of resource competition (e.g., Tillman model) and predation (e.g., Lotka-Volterra model) should be explored to develop the best strategies for working towards the sustainable enterprise. Furthermore, one should take into account the joint optimization of environmental, economic and social dimensions of sustainability.

Concluding Remarks ? An environmentally-challenging industry such as secondary lead smelters requires a fresh look to chart a path towards sustainable development (i.e., survivability and purposive needs) for all stakeholders (i.e., industrial enterprise, individual stakeholders, and social/ecological systems). Such a path should deal with issues beyond pollution prevention, product stewardship and clean technologies.

# Sustainability: Is There a Payoff?


## Authors:

Carberry, J. B., Dupont



Greener facilities and products are a critical part of sustainability, albeit not the full requirement. Identifying and dealing with issues to be ?greener? at manufacturing facilities and in products will be reviewed. Concepts for emerging issues of ?greener? and broader sustainability issues will be presented.

**Life Cycle Assessment (LCA)– A Means to Optimise the Structure of Sustainable Industry**

Narodoslawsky, M., Institute for Resource Efficient and Sustainable Systems
Originally presented on: 8/12/2009 8:30:00 - 9:05:00

Sustainable development will become a forceful driver of re-structuring industry in the 21st century. As a comprehensive development concept, sustainability will influence all aspects of industrial processes, from their raw material base to their size and location to their interaction with each other and the environment and finally to their economic and social implications. This tall task requires completely new approaches to decision support. The methodology for the decision making process within the framework of sustainable development is still far from complete. LCA will, however, almost certainly become part of this process. This contribution explores the capacity of LCA for (and the requirements for LCA methods necessary to) solve questions of optimal process size, raw material base and structure. It will do so by using first generation bioethanol processes and the interaction between an industrial and a municipal energy system as case studies, using the Sustainable Process Index as an LCA evaluation method.

Size and raw material

The first part of the contribution will explore different structural approaches for the production of bio-ethanol, from conventional central bio-ethanol plants with natural gas as an energy source to integrated, more de-central solutions based on various possibilities to provide process heat on the base of renewable resources. This analysis shows that with a change of the raw material base towards renewable resources a completely new discourse about size, location and technology choices will be opened. Besides the well known concept of ?economy of scale? the new concept of ?ecology of scale? will become increasingly important. The analysis shows, that an integrated, de-central process network based on renewable resources will reduce the ecological pressure of providing alternative fuels dramatically.

Integration of industry and societal needs

The second part of the contribution discusses new structural ways to integrate industrial production and societal demand for energy provision. Including societal needs (provision of heating for a city) and changing the raw material base to renewable resources again shows a very high potential for reducing the overall environmental burden while generating positive societal results (jobs as well as regional added value).

New industrial structures and requirements for LCA methods

The case studies show on the one hand that sustainability requires a profound re-structuring of industry as not only economic optimisation will shape industrial processes. The integration of environmental and social considerations into the decision making process will affect the raw material base, size and structure of the industry in the 21st century. As a rule process industry will be faced with the necessity to optimise not only single production lines but more complex, network-like structures, where de-central and central facilities will operate and provide services in terms of energy provision and regional value added besides the provision of goods.

LCA will play a prominent role in the optimisation of such integrated industrial systems. In order to support decisions in future, LCA evaluation methods have to fulfil certain requirements: they must be able to distinguish between different raw material and energy systems (especially to distinguish between fossil and renewable resources) and they must provide easy to read information for decision makers (preferably via aggregated measures). On top of that they must be able to evaluate networks of technologies rather than single technologies and provide clear indication about ecological ?hot spots? within technology networks as well as information about the effect of logistics and transport on the overall ecological pressure of industrial systems.

Key words: LCA, Sustainable Process Index (SPI), sustainable industrial systems, renewable resources

# Life Cycle Assessment at GE: Strategy and Application

## Authors:

Flanagan, W., GE Research

GE's Global Research Center has launched an Ecoassessment Center of Excellence that includes technical expertise in life cycle assessment. The team works with the GE businesses, Global Research, and Corporate Environmental Programs on a variety of product-focused environmental issues. This presentation will focus on GE's evolving vision for LCA as a support tool for new product development and improvement. In addition to comprehensive life cycle assessments, the team is working to implement simplified life cycle tools and processes to guide its internal product development teams. Successes and challenges in these areas will be illustrated with specific examples.
Our approach at the Ecoassessment Center of Excellence is to apply LCA methodologies strategically and selectively. The drivers for LCA within GE are diverse, yet share a common thread: we tend to focus on business-driven product and technology scenarios that require a deeper understanding of complex environmental metrics. A major focus of this presentation will be to explore the drivers and business-context for LCA within GE.

Our LCA efforts support selected company activities in a variety of functional areas including business development, research & development, product development, product evaluation, and commercial support. Illustrative project examples will include alternative liquid fuels, advanced batteries, thin-film solar, disposable products for the biopharmaceutical manufacturing industry, and a pilot Design for Environment activity for medical equipment components.

Our quantitative LCA project execution experience is leading us in new directions as we begin to explore a broader suite of LCA methodologies ranging from comprehensive LCA to streamlined and qualitative LCA techniques. These latter approaches, along with Design for Environment concepts, can be potentially useful for screening and directional guidance, and to expand the application of life cycle concepts beyond the limited reach of expert LCA teams. While LCA is not a panacea, this presentation highlights the benefits and limitations of LCA within a dynamic and innovative business environment.

**Life Cycle Optimization Methods for Enhancing the Sustainability of Design and Policy Decisions**

Keoleian, G., University of Michigan

Originally presented on: 8/12/2009 9:35:00 - 10:00:00

A critical question regarding the life cycle design and management of any product system is, ?What is its optimal service life?? The Center for Sustainable Systems at the University of Michigan has developed life cycle optimization (LCO) methods and models to evaluate optimal service life and asset management decisions from energy, emissions, cost and policy perspectives. This LCO model is based on a dynamic programming method with inputs derived from life cycle assessment (LCA). From an environmental perspective, this is a particularly complex question to resolve for product systems with nonlinear use phase burdens and given uncertainty in technology improvement trajectories. In the case of automobiles, household appliances, and computers, there exist multiple tradeoffs between maintaining an existing model and replacing it with one that is more efficient. The efficiency gain from model replacement should exceed the additional resource investments required to produce the new model.

The life cycle optimization model has also been applied to infrastructure systems that require large capital investments with maintenance costs and have long service lives. Road pavement poses significant modeling challenges given the interactions between pavement and vehicle systems. Models are required to simulate congestion related to road construction events, road deterioration behavior, road roughness effects on fuel economy, and vehicle technology improvements. Here optimization is used to determine asset management decisions including budget allocation decisions, pavement material selection and the timing and frequency of rehabilitation events.

Optimization is an important technique to add to the life cycle modeling toolkit for informing design and policy decisions which will be demonstrated with several applications. LCO has been applied to study automobiles, refrigerators, clothes washers air conditioners and highway infrastructure. Results from these studies will be highlighted and general principles for enhancing sustainability will be presented.

**Managing Strategic Engineering Assets in the 21st Century: A Life Cycle Perspective to Manage Risks**

Liyanage, J., University of Stavanger
Badurdeen, F., University of Kentucky
Originally presented on: 8/12/2009 10:00:00 - 10:25:00

Industrial facilities/assets worldwide have been in continuous change process over a number of years. The ongoing transformations are often attributed to technologically-aggressive, customer-oriented, and market-intensive strategies, where profit margins often remain at the centre of focus of business operations. With the current concerns on more global scale and complex operational conditions, the conventional drivers of competitive advantage have largely been challenged. Besides, network-based, knowledge-intensive societies and partnership-based cooperative economies, have already re-defined asset management excellence. This implies that the success relies heavily on a different set of attributes of business exposure and external demands, and thus the need to be more cautious about novel set of risks in making business decisions. In fact ?now' is the time to break the tradition and to plan for the future to mitigate hidden business risks in a complex, vulnerable, and a rapidly changing environment. The questions are then, what are these hidden risks, what issues are critical for strategic assets to mitigate those hidden risks, how to prepare for the future through a risk-aversive value-creating roadmap, how to operationalize forward-looking actions, and what can be learnt from the best players in other industries to secure long-term growth while maintaining good profitability.

This paper brings these issues into discussion in detail through various examples, challenging and setting a road-map for strategic asset management to identify the hidden risks and to capitalize on the value-potential of complex business activities in a sustainable perspective.

Biosketch: Dr. Liyanage is a Professor of Industrial Asset Management at the Faculty of Science and Technology, University of Stavanger, Norway. He earned his Ph.D. in Offshore Engineering (specializing in Asset Operations and Maintenance) from the Norwegian University of Science and Technology in 2003. Dr. Liyanage has done extensive work in the sustainable asset management area, collaborating with several companies in Norway (for example joint projects with Conco Phillips, Eni Norway, Petrolink, Gassco, AGR, Seimens Turbomachinery AB). He is an honorary founding fellow of the International Society for Engineering Asset Management (ISEAM), a member of the International Federation of Automatic Control (IFAC) special working group on ?Advanced maintenance engineering, Services, and Technology, Society of Petroleum Engineers (SPE) and an honorary fellow of the International Foundation for Research in Maintenance (IFRIM). He is the founding editor-in-chief of the International Journal of Strategic Engineering Asset Management and is an editorial board member of the International Journal of Sustainable Manufacturing. He has published extensively in reputed international journals, delivered keynote speeches and presented at numerous national and international conferences.

# Atmospheric Pressure Plasma Treatment: Enabling Environmentally Friendly Surface Treatment Processes

## Authors:

Glogauer, S., Plasmatreat North America

This paper will outline the basic premise of surface treatment requirements, introduce the concepts of plasma and plasma treatment, describe the effects of plasma treatment, detail the environmental benefits of plasma via case studies and end with a short summary. A quick introduction of current surface treatment methods will be discussed and compared with plasma treatment. Competing technologies include flame and corona treatment. A short overview of plasma and how it is generated and used in an industrial setting will be explained. Atmospheric pressure plasma (as opposed to low pressure or vacuum plasma) is created by exposing a process gas to a high frequency, high voltage electric discharge that excites electrons and induces them to leave their orbits, creating simultaneously free electrons and ions, and thus a plasma. The effect of plasma treatment on substrates and bonding parameters will then be discussed. This treatment must produce clean and highly activated surfaces with reliable and repeatable characteristics in order to achieve optimal adhesive bonding, coating, and printing results. The remainder of the paper will focus on the elimination of environmentally unfriendly surface treatment processes such as: solvent wiping, power washing, and the application of adhesion promoters and primers. These will be demonstrated through the use of several case studies including the potential elimination of both power washing and application of adhesion promoter to plastic parts prior to paint applications, as well as the elimination of harsh cleaning chemicals prior to coil coating processes. Plasma treatment can not only eliminate wet chemistry applications, but also improve adhesion properties and the quality of finished parts, reduce energy consumption, and streamline the manufacturing process resulting in both environmental and cost benefits.

**Using LCA to Measure Sustainability**

Laurin, L., EarthShift
McEwen, L., EarthShift
Originally presented on: 8/12/2009 11:10:00 - 11:35:00

LCA has the unique distinction of being the only scientifically-based, comprehensive method to measure ?green.? While LCA should be augmented with environmental risk assessment and some measure of social assessment, it can be the cornerstone of a company's sustainability assessment.

But what is sustainability? If our definition is something like the Report of the World Commission on Environment and Development:

?Sustainable development is development that meets the needs of the present without compromising the ability of future generations to meet their own needs.?

it becomes clear that no company or product in today's environment is sustainable, except perhaps a few subsistence farmers in remote corners of the world. We are on a journey to sustainability, we're not there yet.

How do we assess products to get from here to there most quickly? Early studies of photovoltaic panel manufacture showed that panels would have to work at peak efficiency for as much as 6 years to create as much energy as was used to make the panels in the first place. (Corkish, 1997). Did this heavy upfront energy investment mean that we should have abandoned research in this area? Studies of reusable drinking cups show that the hot water used for washing brings the impacts of these cups close to single use cups (OVAM, 2006). Does this indicate that we should continue work to make the single use cups even less impactful, or is there another alternative? This presentation will explore several scenarios through possible future paths of development with the aim to broaden the discussion of the path to sustainability and the care needed to be sure we don't abandon the most viable paths before their true potential can be understood.

References:

Corkish, Richard, ?Can Solar Cells Ever Recapture the Energy Invested in their Manufacture?? Solar Progress, (Australia and New Zealand Solar Energy Society) vol. 18, No. 2, pp. 16-17, 1997

OVAM , ?Comparative LCA of 4 types of drinking cups used at events,? OVAM, 2006

**The Sustainable Environmental Performance Indicator: LCA Based Strategic Decision Making**

De Benedetto, L., University of Pannonia
Klemes, J. J., University of Pannonia
Originally presented on: 8/12/2009 11:35:00 - 12:00:00

Life Cycle Assessment (LCA) is a well used tool for analyzing environmental impacts on a wide perspective and with a reference to a product system and economic activity. However it has some limitations and there is a need for a novel approach that complements environmental and financial considerations. This is addressed in this study with the introduction of a new graphical representation: the Environmental Performance Strategy Map. This particular graphical map allows combination of the main environmental indicators (footprints) with the additional dimension of cost. Into a Sustainable Environmental Performance Indicator.

## 1. From Environmental Assessment to Strategic Environmental Map

The ecological footprint is a way to compare human demand with our planet capacity to regenerate it and it is measure of our burden on the ecosystem. In a broader view the ecological footprint is related to the method of LCA. One of LCA advantages is that it better covers the whole range of impacts, and it may also provide an accounting of the upstream impacts. Nevertheless, one of the most important limitations in the application of LCA as an input for strategic decision-making is the limited inclusion of cost and investment considerations. A new approach is required to integrate financial, environmental, resource and toxicological considerations into a single analysis. The core of the concept is to calculate some specific sustainability indicators based on LCA. The cradle to grave approach will assure that all environmental and human consequences taken into account. To represent these relations and to compare options from an environmental and, more generally, business perspective a new graphical representation needs to be introduced: the Environmental Performance Strategy Map (EPSM). The objective of this representation is to build upon the strength of Ecological Footprint and Life Cycle Analyses to provide a single indicator for each option. The first step in building the EPSM correctly is to calculate the impact of the option under analysis for all of the main environmental burdens. The combination of these elements and the cost perspective will provide a single indicator to assign to each option.

## 2. Selection of Footprints

Different methods have been developed in the last years to correlate environmental sustainability of specific activities with land and water areas required to supply this activity with resources and to absorb its wastes (Monfreda et al. 2004). This is usually referred to as Ecological Footprint. The difficulty in using the tool in the decision making process have been noted by Ayres (2000) and overcome by the development of specific indicators - SPI (Krotscheck and Narodoslawsky 1996) and DAI (Eder and Narodoslawsky 1999). In particular the Sustainable Process Index (SPI) considers the area as a basic measure: the more area a process requires, the more its burden from an ecological point of view.

To provide a more comprehensive analysis of the interaction of the environmental burdens and financial costs newly developed the EPSM is based on the combination of various footprints (further details are available in De Benedetto and Kleme? 2009):

- Carbon footprint (Hujbregts et al. 2008, Wiedmann and Lenzen 2007)
- Water footprint (Hoekstra and Hung 2002, Hoekstra 2007)
- Energy footprint (Land, Renewables, Non-Renewables) (Stoeglehner 2003)
- Emission footprint (emissions in Air, in Water, in Soil, Waste materials) (Sandholzer and Narodoslawsky 2007)
- Work environment footprint (work-environment and toxicological impacts) (Schmidt et al., 2004)
-

3. Building the Map

Once the contribution of each option to the specific footprints has been calculated, it is possible to build the EPSM. The basic concept is to map the footprint on a specific spider-web plot (Fig. 1, a). To compare measures the results of each footprint are normalized, resulting in a scale from 0 to 100. A deviation-from-target methodology is proposed, where for each of the footprints, the targets are either based on maximum available resources or are drawn from scientific consensus or regulatory requirements. Each value is recorded as a percentage. The aim is to lower as much as possible the contributions of each footprint. Each option has an area assigned that represents a combination of all footprints. To specify the cost and financial impacts an additional dimension has been introduced (Fig.1, b). The indicator takes into account the total financial investment required for each options. The volume of each pyramid represents the overall environmental and financial impact. The authors define this index as the Sustainable Environmental Performance Indicator (SEPI).

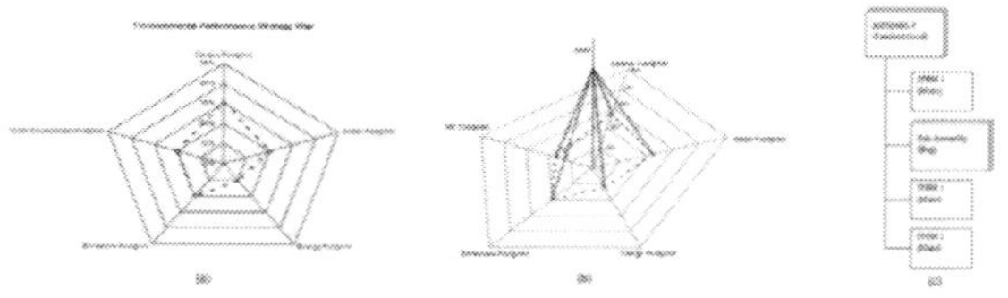

Fig. 1 (a) Plotting the footprints in the EPSM , (b) The additional dimension in the Environmental Performance Strategy Map, (c) Example of simplified Bill of Materials

Finally it is possible to plot all options under consideration in a specific EPSM. The map thus enables comparison of different options for strategic decision-making purposes, based on a single sustainability indicator.

4. The Environmental Bill of Materials and Technology Routing

The EPSM considers production processes as a whole. In case a more granular approach is required we need to introduce two new elements. The term ?Bill of Materials? (BOM) is used to indicate basic materials, components, parts and the quantities of each needed to manufacture a product or service (Reid and Sanders 2002). The idea is to associate contributions to the above mentioned footprints to each component of a manufactured good or service (Fig 1, c). With regards to the previous work on this subject (De Benedetto and Kleme? 2009), the contributions of each item to the footprints specified in the previous paragraph is recorded (Environmental Performance Points or EPP). We define this new BOM as Environmental Bill of Materials (ENV-BOM).

| ENV-BOM | | | | Environmental Performance Points | | | | |
|---|---|---|---|---|---|---|---|---|
| Assembly | Type | UOM | Qty | Carbon | Water | Energy | Emissions | Work Env. |
| Item 1 | Make | Each | 1 | | | | | |
| Sub-Assembly | Buy | Each | 1 | | | | | |
| Item 2 | Buy | Each | 1 | | | | | |
| Item 3 | Make | Each | 1 | | | | | |

Table 1. Example of Environmental BOM

When all components are recorded it is possible to consider also the operations and the technology process. We define the collection of all operations Technology Routing (Table 2). As done for the BOM, it is necessary to define the contributions in terms of the footprints for each step of the process. Table 2. Example of Technology Routing table:

| Technology Routing | | Environmental Performance Points | | | | |
|---|---|---|---|---|---|---|
| Operation | Quantity | Carbon | Water | Energy | Emissions | Work Env. |
| Manual Assembly | 1 | | | | | |
| Painting | 1 | | | | | |
| Drying | 1 | | | | | |
| Quality Control | 1 | | | | | |
| Packaging/Shipping | 1 | | | | | |

When all Environmental Performance Points are defined, it is possible to sum them all to provide the final figures to build the Environmental Performance Strategy Map.

## 5. Conclusions

The Sustainable Environmental Performance Indicator, and the associated Strategy Map, provide an overall indicator of the environmental performance. One of its main characteristics is that it aggregates all contributions in a very high level. In case this is not desirable and a more granular approach is required, we introduced two new elements: the Environmental Bill of Materials and Technology Routing. These elements allow the decomposition of each environmental burden at materials and process steps level. This approach has been demonstrated with specific case studies.

Acknowledgement:

Financial support from MC Chair (EXC) INEMAGLOW: FP6-042618 Integrated Waste to Energy Management to Prevent Global Warming is gratefully acknowledged.

References

Ayres R.U. 2000, Commentary on the utility of the ecological footprint concept, Ecol. Econ, 32, 357-358

De Benedetto L. and Kleme? J., 2009, The Environmental Performance Strategy Map: an Integrated LCA approach to support the Strategic Decision Making Process, J Cl Prod, in press

Eder P., Narodoslawsky M., 1999, What environmental pressures are the region's industry responsible for? A method of analysis with descriptive input, output models. Ecol. Econ., 29, 359-374

Hoekstra A.Y. and Chapagain A.K., 2007, Water footprints of nations: water use by people as function of their consumption pattern. Water Res. Mgmnt., 21, 35-48

Hoekstra A.Y. and Hung P.Q., 2002, Virtual water trade: a quantification of virtual water flows between nations in relation to international crop trade. Value of water research series, 11

Hujbregts M.A.J., Hellweg S. and Frischknecht R. et al., 2008, Ecological footprint accounting in the life cycle assessment of products. Ecol. Econ., 64,798-807

Krotscheck C. and Narodoslawsky M., 1996, The Sustainable Process Index. A new dimension in ecological evaluation. Ecol. Eng., 6, 241-258

Monfreda C., Wackernagel M. and Deumling D., 2004, Establishing natural capital accounts based on detailed ecological footprint and biological capacity assessment. Land Use Policy, 21, 231-246

Reid, R. Dan; Sanders, Nada R. (2002) Operations Management John Wiley & Sons 457?458. ISBN 0-471-32011-0

Sandholzer D. and Narodoslawsky M., 2007, SPIonExcel?Fast and easy calculation of the Sustainable Process Index via computer. Res. Cons. Recycl., ,50, 30-142

Schmidt A., Poulsen P.B., Andreasen J., Floee T. and Poulsen K.E., 2004, The working environment in LCA. A new Approach. Danish Environmental, 72.

Stoeglehner G., 2003, Ecological footprint ? a tool for assessing sustainable energy supplies. J. Cle.Prod.,11,267-277

Wiedmann T. and Lenzen M., 2007, On the conversion between local and global hectares in ecological footprint analysis. Ecol. Econ., 60, 673-677

# Sustainable Design of Chemical and Biochemical Processes: The Role of Models and Modelling

## Authors:

Carvalho, A. I. C. S. G., Instituto Superior Técnico

Models play a very important role in the sustainable design of chemical and biochemical procsses. The design related calculations use models of different scales, from different sources and of different types. The reliability, consistency and applicability of the models are issues that need to be addressed by methods specially developed for sustainable design of processes.
A systematic methodology for sustainable process design and a corresponding software (SustainPro) has been developed by the authors. This methodology can be applied for sustainable design and analysis of chemical processes operating on batch as well as continous operation modes. A computer-aided framework has been designed and implemented to allow the integration of models and data from various sources with the work-flow and data-flow corresponding to sustainable design of processes. At each step of the work-flow, model-based methods and tools are applied with their corresponding data-flow requirements.

The methodology (and the software, SustainPro) first performs a flowsheet decomposition based on the identified component mass and energy ?paths?. Then it calculates a set of mass and energy indicators that trace the paths of the components ?mass-flows? and ?energy-flows? as they enter (or generated) and leave (or consumed) the process. With this information, the critical points of the process are identified, followed by a sensitivity analysis to locate those critical points which have the potential for effecting the most sustainable improvements in the process. Through this sensitivity analysis the target indicators are selected and once the target values are established, another sensitivity analysis is performed to determine the design and the operational process variables that most influence the targets. The next step involves the matching of the defined targets with process designs in terms of changes in the sensitive design and operational process variables. This last step is performed using a set of algorithms for process synthesis that allow the generation of feasible flowsheet or operational alternatives. In order to determine if the generated alternatives are better than the reference design, a set of sustainable metrics and safety indices are used as performance criteria.

from the above description, it is clear that models and/or data are used in almost all steps of the methodology. For example, process simulation models (and corresponding data) are employed to establish the reference design and for analysis of design alternatives. Indicator-models are needed to identify process points that can be targeted for improvements. These models use process simulation data, cost data as well as physical properties data. For evaluation of sustainable design alternatives, models for metrics (sustainability and safety) are needed. Also, models and associated data for LCA would be needed if this option is to be used. Finally, for generation of design alternatives, model-based process synthesis methods are needed. Note that almost all the models employed at various steps of the design work-flow, employ parameters that are usually obtained from different sources. The models need to be qualitatively correct during the synthesis steps while they need to be quantitatively correct during the final design calculations.

The presentation will highlight the work-flow and data-flow for the developed systematic methodology for sustainable process design together with the models used at each step. The need and use for a computer aided modelling tool to quickly generate, analyze and adopt models for the various steps of the design work-flow will be illustrated. The uncertainty of model parameters and their effect on design decisions will be highlighted through two case studies operating in batch and continous modes of operation.

**Applications of P-Graphs for Enhancing Sustainability of Industrial Plants**

Friedler, F., University of Pannonia
Varbanov, P. S., University of Pannonia
Fan, L. T., Kansas State University

Originally presented on: 8/12/2009 9:05:00 - 9:35:00

## Introduction

The consideration of sustainability in the design and operation of industrial processes is becoming a key issue of process systems engineering. There are Mathematical Programming (MP) tools for designing chemical processes by both maximizing the profit and minimizing the costs. These MP formulations are usually highly complex in solving practical problems. Therefore, the original problems are usually decomposed into smaller ones, solved sequentially.

Unfortunately, these MP-based approaches cannot be adopted for designing sustainable processes for at least two reasons: (i) the consideration of sustainability makes the originally complex problem even more complex; (ii) the decomposition of the model (and problem) may produce inadequate results. Therefore, for designing sustainable processes, the methodology must be highly effective and capable of integrating the subproblems of the overall design problem.

## The issue of complexity

The complexity of synthesizing process systems and those for energy recovery in particular, in the form of flowsheets, arises from the simultaneous involvement of continuous as well as discrete entities. The combinatorial complexity increases exponentially with the number of candidate operating units. The larger the number of candidate units for the system, the greater the number of possible combinations. MP has achieved only moderate success in reducing the search space, and therefore, the deployment of the approach based on MP becomes increasingly difficult as the size of problems enlarges: (i) as the size of an algebraic optimization problem grows, it is destined to become impossible for any available solver to examine the combinations of the values of integer variables from the topological point of view; and (ii) an enormous number of topological options makes it rather difficult to build the superstructures necessary for solving the problem. Consequently, practical problems are too complex to be solved. If the problem is excessively simplified so that it is solvable by MP, the resultant problem may no longer adequately represent the original one.

## Process optimization applications - energy conversion and recovery

The extent of climate change depends on the concentrations of $CO_2$ and other greenhouse gases in the atmosphere. From the energy perspective, therefore, the enhancement of the industry's sustainability mainly necessitates the reduction of $CO_2$ in terms of emission from manufacturing. Three major means are available for reducing $CO_2$ emissions. These include improving energy conversion efficiency, increasing the $CO_2$ recycling via the use of biofuels, and $CO_2$ sequestration (Kleme? et al., 2007; Varbanov et al., 2008).

The first area of interest is elevating the efficiency for power generation, especially through combined heat and power (CHP) generation. This can be achieved by incorporating fuel cells in combined (hybrid) cycles. The fuel cells play the role of the topping cycle, whereas the bottoming cycles can be steam or gas turbine subsystems. Various operating modes are possible to achieve increased energy conversion efficiency as well as wide ranging power-to-heat ratios.

The second area of interest is the design of energy efficient processes. Many processes, e.g., distillation and evaporation often constituting the core of any process system, are energy intensive, thus requiring large amounts of energy. Heat integration has been the preferred method of achieving savings of utilities in addition to the reduction in the energy intensity of the core processes.

The classes of problems in the two areas, mentioned above, lead to process-synthesis formulations involving network optimization, which is characterized by significant combinatorial complexity. Initial systematic approaches to circumventing such complexity resorted to a combination of thermodynamic targeting and a heuristic design procedure by means of Process Integration (PI) or Mathematical Programming (MP). The latter represents the selection of the operating units via integer variables. Flexibility requirements have been added to the design problems by recognizing the inherently time-varying nature of process operation due to changing ambient conditions, worker shifts, and market supply-demand balances. The multi-period optimization paradigm has emerged from the methodologies for synthesizing flexible HENs based on an MILP model; see Floudas and Grossmann (1986). MP has also been applied to this problem by Aaltola (2002). Ahmad et al. (2008) have adopted Simulated Annealing.

**P-graph framework**

The P-graph framework has been developed for systematic design of industrial processes (Friedler et al., 1992, 1995, 1996) to maximize the profit or minimize the costs. This approach proved to be effective in solving industrial-scale problems. Later, it has been extended to the simultaneous solution of formerly sequentially solved problems, e.g. simultaneous synthesis of a process and its heat exchanger network (Nagy et al., 2003), and simultaneous synthesis of separation and heat exchanger networks (Heckl et al., 2005). In the present work the P-graph framework has been extended to consider sustainability indicators in designing industrial processes.

Major challenges and limitations of MP for multi-period HEN synthesis can be effectively overcome by resorting to P-graphs (process graphs). The present work explores the application of an advanced approach to the HEN synthesis based on hP-graphs rooted in the P-graph framework. This framework offers a highly viable alternative to the MP framework in circumventing the combinatorial complexity of the problems by judiciously exploiting their network characteristics.

## References


Aaltola J., 2002. Simultaneous synthesis of flexible heat exchanger networks. Appl Therm Engng, 22(8):907?918.

Ahmad M.I., Chen L., Jobson M., Zhang N., 2008. Synthesis and optimisation of heat exchanger networks for multi-period operation by simulated annealing. Proceedings of PRES2008/CHISA2008, Prague, 4:1200

Floudas C.A., Grossmann I.E., 1986. Synthesis of Flexible Heat Exchanger Networks for Multiperiod Operation. Comp Chem Engng, 10(2):153-168.

Friedler F., Tarjan K., Huang Y.W., Fan L.T., 1992. Graph-Theoretical Approach to Process Synthesis: Axioms and Theorems. Chem. Eng. Sci., 47(8):1972-1988.

Friedler F., Varga J.B., Fan L.T., 1995. Decision-Mapping: A Tool for Consistent and Complete Decisions in Process Synthesis. Chem. Eng. Sci., 50, 1755-1768.

Friedler F., Varga J.B., Fehér E., Fan L.T., 1996. Combinatorially Accelerated Branch-and-Bound Method for Solving the MIP Model of Process Network Synthesis. In State of the Art in Global Optimization, Ed. Floudas, C.A. and Pardalos, P.M., Kluwer Academic Publishers, Boston, Mass, 609-626.

Heckl I., Friedler F., Fan L.T., 2005. Integrated Synthesis of Optimal Separation and Heat Exchanger Networks Involving Separations Based on Various Properties, Heat Transfer Engineering, 26(5), 25-41.

Kleme? J.; Bulatov I., Cockeril T., 2007. Techno-Economic Modelling and Cost Functions of CO2 Capture Processes. Computers & Chemical Engineering, 31(5-6):445-455.

Nagy A.B., Adonyi R., Halasz L., Friedler F., Fan L.T., 2001. Integrated Synthesis of Process and Heat Exchanger Networks: Algorithmic Approach. Appl. Therm. Engng, 21:1407?1427.

Varbanov P., Friedler F., 2008. P-graph Methodology for Cost-Effective Reduction of Carbon Emissions Involving Fuel Cell Combined Cycles. Applied Thermal Engineering, 28(16): 2020-2029.

**Sustainable Carbon Footprint Reduction by Integrating Renewables Into Total Sites**

Varbanov, P. S., University of Pannonia
Klemes, J. J., University of Pannonia
Originally presented on: 8/12/2009 9:35:00 - 10:00:00

The majority of industrial, residential, service and business, agriculture farms and production settlements energy systems are still dominated by the fossil fuels as primary energy sources. They are mostly equipped with steam and/or gas turbines, steam boilers and water heaters (running on electricity or gas) as energy conversion units. The challenge to reduce the Global Warming impact can be efficiently addressed by increasing the share of renewables in the primary energy mix, leading to smaller carbon footprint of the related activities. This requires integrating solar, wind, biomass, as well as some types of waste together with the fossil fuels.

The current work analyses integrating renewable heat into Total Sites comprising industrial processes, residential areas, building complexes (hospitals, universities, hotels, business centres) and farms. The variability in energy demands and supplies are taken into account.

## 1. Introduction

Renewable resources are usually available on smaller scale distributed over a given area. Their availability, with the exception of biomass, is usually well below 100 %. The resource availability varies significantly with time and location. This is caused by the changing weather and geographic conditions. The energy demands (heating, cooling and power) of the considered sites vary significantly with time of the day and period of the year.

Therefore, designing energy conversion systems for utilisation of renewable resources is more complex than when using just fossil fuels. By combining the supply and demand streams of the individual users, such systems may serve industrial plants as well as residential customers and the service sector (hotel complexes, hospitals).

A novel approach has been extending the Heat Integration methodology to including waste and renewables (Perry et al. 2008) for a given steady state of supply and demand.

## 2. Issues and concepts for resolving them

### 2.1. Combining consumer sites and resources

The energy demands vary with the types of end users as well as with the time schedules. However, whenever it is geographically and safety- feasible, the simultaneous integration of Industrial sites, residential areas and service buildings with renewable energy sources can bring significant benefits. The main reasons for this are that the application demands vary in different ways, potentially complementing each other, as well as the possibility for inter-process heat integration.

### 2.2. Demand characteristics

Both the supply and demand for energy vary with time and location. To simplify the initial analysis, a given fixed a set of locations is assumed.

The time variations of energy demands have been subject to research both in industrial and residential contexts. An example is a study investigating the variation of residential energy consumption for heating, electricity and hot water (Bance, 2008). The results show two types of trends ? hourly variations during each day and seasonal variations during the year. For the hourly variations there are nearly steady periods during the usual office hours and two consumption peak intervals in the morning and in the evening. The seasonal variations are relatively smooth with more substantial space heating demands from October until April.

For their efficient exploitation it is necessary to assess renewables' overall availability and variability with time. Some of them are close to the performance of fossil fuels and can be well stored for continuous energy generation. An example is biomass, where the supply varies by year seasons and by bio-waste availability. However, sufficient storage could be made available. The availability of other renewable sources as wind and solar varies faster ? in hours and even minutes.

These types of variation present an integration challenge where the time horizons of the changes are diverse. From the given examples, for biomass, the time slices needed would last on the order of months and at smallest ? weeks. For wind and solar energy, the time slice durations will obviously be much shorter. This brings a necessity to extend the Total Site methodology (Dhole and Linnhoff, 1993; Kleme? et al., 1997) to deal with the described variations.

### 2.3. Integrating the total sites including renewables

The Heat Integration methodology has been traditionally dealing with industrial plants where most effort has been made to achieve the stable steady-state production (Linnhoff and Hindmarsh, 1983). It was summarised by Smith, (2005) and more recently by Kemp (2007). The second step has been to integrate more industrial plants into Total Sites (Dhole and Linnhoff, 1993; Kleme? et al., 1997), where more than just industrial plants may be included (Perry et al. 2008).

The maximum availability of renewables is usually limited and cannot be controlled by the energy conversion system. To integrate renewables as fixed and not belonging to the degrees of freedom is obviously not correct. What can be used as a degree of freedom is the fraction of the renewable resources to be harvested, compared with their overall availability.

## 3. Integration approach

To account for the variation of the demands, the renewables availability and simultaneously maximising the heat recovery, it is necessary to apply Total Site integration and when needed to consider heat storage possibilities. This is performed by constructing Total Site Profiles and Total Site Composite Curves for a set of time intervals, referred to as ?Time Slices? and maximising the heat recovery within each Time Slice. This formulation extends the methodology developed previously for batch processes (Kleme? et al., 1994; Kemp and Deakin, 1989).

A very important issue related to the system design is the need for heat storages. If this is available at given time and required capacity with feasible cost it could considerably increase the system efficiency. The energy storage is complicated and demanding issue, which is still waiting for major breakthrough. The Heat Integration methodology could contribute to this problem solution by providing targets which are supposed to be achieved.

## 4. Conclusions

The inclusion of renewables with their changing availability requires extensions of the traditional heat integration approach. The problem becomes more complicated and has several more dimensions. Revisiting some previously developed Process Integration tools and their further development enables solving this extended problem. The presented contribution has been a step in this direction summarising the problem and suggesting some options for its solution. A demonstration case study illustrates the heat saving potential of integrating various users and using heat storage. The advanced tools based on the suggested methodology have been under development.

## References


Kleme? J., Linnhoff B., Kotjabasakis E., Zhelev T.K., Gremouti I., Kaliventzeff B., Heyen G., Maréchal F., Lebon M., Puigjaner L., Espuña A., Graells M., Santos G., Prokopakis G.J., Ashton G.J., Murphy N.,. Paor de A.M, Kemp I.C., Design and operation of energy efficient batch processes, Final Report, Commission of the European Communities. Brussels, 1994.

Kemp, I., C., Deakin, A., W., The Cascade Analysis for Energy and Process Integration of Batch Processes. Part 1: Calculation of Energy Targets, Chem. Eng. Res. Des., vol. 67, pp.495-509, 1989.

Perry, S., Kleme?, J., Bulatov, I., Integrating Waste and Renewable Energy to Reduce the Carbon Footprint of Locally Integrated Energy Sectors, Energy, vol. 33, no. 10, pp. 1489-1497, 2008.

Bance, P., Residential-Scale Fuel Cell CHP: A Better Match for Domestic Loads, Cogeneration & On-Site Power Production, vol. 9, no. 3, Available from www.cospp.com/display_article/330132/122/CRTIS/none/none/1/Residential-scale-fuel-cell-CHP-a-better-match-for-domestic-loads/, Last accessed on 04.07.2008, 2008.

Dhole, V., R., Linnhoff, B., Total Site Targets for Fuel, Co-Generation, Emissions, and Cooling, Computers and Chemical Engineering, vol. 17, Supplement, pp. S101-S109, 1993.

Kleme?, J., Dhole, V., R., Raissi, K., Perry, S., J., Puigjaner, L., Targeting and Design Methodology for Reduction of Fuel, Power and $CO_2$ on Total Sites, Applied Thermal Engineering; vol. 7, pp.993?1003, 1997.

Linnhoff, B., Hindmarsh, E., The Pinch Design Method for Heat Exchanger Networks, Chemical Engineering Science, vol. 38, pp. 745-763, 1983.

Smith, R., Chemical Process Design and Integration, John Wiley and Sons Ltd., 687 pp., ISBN: 0471486809, 2005.

**Design Tools for Sustainable Water Transmission and Distribution**

Miller, C. J., Wayne State University
McElmurry, S. P., Wayne State University
Originally presented on: 8/12/2009 10:00:00 - 10:25:00

Water utilities represent a significant fraction of the total energy demands of a typical community. In the case of large urban centers with limited storage capacity, the water utility may be the primary energy consumer. The use of optimized pump operation strategies in these systems may achieve significant energy savings. Although some utilities have an energy optimization program, those currently available are not integrated with real-time data from the plant. Instead, the programs are typically used on an intermittent basis to specify pump settings for future expected loads/demands. To make a truly integrated energy optimization system, real-time data on system performance, water use and demands, are necessary inputs. Previous technologies rely more on archived demand patterns and one isolated point optimization rather than system-wide optimization for real-time conditions. The key improvement of the present program is the use of real-time data gathered from throughout the system to provide ?on the fly? optimization of the water system. Instead of conducting pumping optimization only for individual pump facilities, the revised program will perform systematic optimization for the utility's entire water transmission and distribution system.

A pilot program based on these concepts is being developed for the Detroit Water and Sewerage Department (DWSD). The pilot program illustrates the use of the data collection system and the input of this data into a real-time optimization program. The primary benefit of the project is a reduction of total energy consumption as well as a reduction in the peak energy needs of water utilities. These reductions will lead to reduced reliance on coal-fired power plants (the primary energy source in the Great Lakes region) and a concomitant reduction in the environmental/ecological impacts of such emissions. The impacts of fossil fuels are well documented and include impacts to personal health, surface waters, soils, vegetation, terrestrial and aquatic life, and aesthetics. Furthermore, the ecological damage associated with ever increasing energy consumption is not limited to those states relying on fossil fuel. Other forms of energy production have their own unique impacts on the ecosystem; these impacts are not limited to traditional energy sources but are becoming more understood for alternative energy technologies (AET) as well.

**Sustainability Under Severe Uncertainty: A Methodological Study**

Lou, H. H., Dan F. Smith Department of Chemical Engineering
Huang, Y., Wayne State University
Zheng, K., Lamar University

Originally presented on: 8/12/2009 10:45:00 - 11:10:00

Sustainable development needs not only the evaluation of the status quo, but also prediction and strategic decision-making for the future. During a methodological study on sustainability, one of the most challenging tasks is how to deal with uncertainty, since the available information and data (technical or non-technical) are almost always uncertain, incomplete, and imprecise, and the possessed knowledge is usually insufficient and imperfect. Clearly, these difficulties hinder the advancement of sustainability engineering and science. In this presentation, the applicability of three types of approaches to handling severe uncertainty: the information gap approach, the probability bounds analysis approach, and the fuzzy logic approach will be analyzed and assessed. A general guideline for handling uncertainties in modeling, analysis, and decision making will be presented. This will be particularly valuable for sustainability-purposed modeling, analysis, design, and decision-making.

# A Modular Approach to Sustainability Assessment and Decision Support in Chemical Process Design

## Authors:

Othman, M. R., Berlin Institute of Technology
Repke, J., Technical University of Berlin

Selection of sustainable process design at early design stage is important to minimize long term effect to the environment and future generations. In the chemical industry, process design sustainability, in terms of energy and material use efficiency, waste minimization, process and product life cycle performance, total cost accounting, etc, is of upmost importance. Over the past decade, a variety of sustainability indicators have been introduced, but with various challenges. It becomes clear that the industry needs urgently practical tools that can be used to conduct convincing systematic sustainability assessment on existing processes and/or new designs, and further to obtain decision support for necessary process modification or selection of design alternatives. This work presents a systematic modular approach for sustainability assessment and selection of process designs through integrating hard (quantitative) economic and environmental criteria along with soft (qualitative) quality criteria into the activities. The approach contains four modules: 1) a process simulator module, 2) an equipment and inventory acquisition (EIA) module, 3) a sustainability assessment (SA) module, and 4) a decision support process (DPS) module. The modules fully utilize the capabilities of process simulators and third party software, such that case model development, problem definition and formulation, data acquisition and analysis, and decision support, that are systematically integrated. The approach can facilitate process designers and decision makers significantly in conducting structured and accurate design assessment and deriving design decisions at early design stage. The functionality of the introduced approach is illustrated by an example of biodiesel process design. Insightful sustainability analysis and persuasive decision support in this case study demonstrate its superiority over commonly practiced techno-economy evaluation approaches.

**Simulation Tools for Design of the Next Generation of Milk Processing Plants**

Tomasula, P. M., USDA-ARS Eastern Regional Research Center
Nutter, D. W., University of Arkansas
Yee, W. C., USDA-ARS Eastern Regional Research Center
Mc Aloon, A., USDA-ARS Eastern Regional Research Center
Originally presented on: 8/12/2009 11:35:00 - 12:00:00

A growing number of food retailers have incorporated environmental sustainability into their business plans and are concerned about the greenhouse gas emissions (GHG) and the carbon footprint associated with the fluid milk value chain. The fluid milk value chain begins on the farm with crop production and then milk production; the raw milk is then transported to the processing plant for processing, packaging, and distribution to retail, where the milk is refrigerated until sold. Estimated sources of GHG for the fluid milk value chain total 28 million metric tons of carbon dioxide/year with approximately 80% of the emissions generated on the farm. A Dairy Sustainability Summit, held in 2008, brought together representatives of the entire fluid milk value chain with the goal of creating initiatives for each sector of the value chain to reduce GHG by 25% by the year 2020. One of the initiatives created at that meeting was the D-CREE Initiative (Dairy Processor Carbon Reduction Through Energy Efficiency). In one of its objectives, a fluid milk process simulation tool is to be developed to allow processors to assess individual plants and their next practice opportunities. Tools are available for estimating overall GHG production but do not provide quantitative information on mass and energy flows or water usage at each unit operation in a processing plant. We have used a commercial process simulator program to generate a flow sheet and calculate the mass and energy usage of each unit operation for small, medium, and large milk processing plants. The associated GHG were also calculated. Our long-term goal is to compare the energy and GHG data with data obtained from processing plants to refine the simulation so that processors can assess their own plants and use sensitivity analysis to test next practice opportunities, such as alternative unit operations and alternative energy technologies that reduce GHG. The application of the simulator to development of sustainable milk processes will be discussed.

**Standards for Biofuels / TheTripartite Review of Performance Standards**

Corr, C., Archer Daniels Midland Company

Originally presented on: 8/12/2009 13:30:00 - 14:10:00

This presentation will review the history of the Tripartite process used to review the performance standards from the European Union, Brazil and the United States. Through the process it became clear that regional performance standards are much more appropriate than a single international standard. The Tripartite experience provides excellent lessons for sustainability standards.

**Development of Biodiesel Standard Reference Materials**

Schantz, M. M., NIST

Originally presented on: 8/12/2009 14:10:00 - 14:35:00

The National Institute of Standards and Technology (NIST), USA and the National Institute of Metrology, Standardization, and Industrial Quality (INMETRO), Brazil are collaborating on the development of Certfied Reference Materials (CRMs) for biofuels. INMETRO is developing two bioethanol CRMs, and NIST is developing two biodiesel CRMs, Standard Reference Material (SRM) 2772 B100 Biodiesel (Soy-Based) and SRM 2773 B100 Biodiesel (Animal-Based). Biodiesel is an alternative fuel produced by the transesterification of triglycerides of vegetable oils or animal fat with methanol to form the fatty acid methyl esters. Since some of the by-products, if not separated, can cause engine problems, it is important to have reliable concentrations for glycerol, mono- and diglycerides, unreacted triglycerides, and fatty acid methyl esters. The biodiesel industry within the USA is increasing at a fast pace, and the industry needs reference materials to benchmark the measurements not only of the fatty acid methyl esters and glycerides but also for sodium, potassium, calcium, magnesium, sulfur, phosphorous, moisture, density, and viscosity. NIST and INMETRO have provided data for the parameters that they developed methods for in these two SRMs. The data have been combined to provide certified concentration values for fatty acid methyl esters, water, sulfur, and kinematic viscosity; reference concentration values for additional fatty acid methyl esters, free glycerol, and additional physical properties; and information values for other trace elements, mono-, di-, and triglycerides, and total glycerol. The development and certification of these biodiesel SRMs will be discussed.

**ASTM Standards as Part of a Comprehensive Approach to Sustainability**

Picariello, P., ASTM International

Originally presented on: 8/12/2009 14:35:00 - 15:00:00

ASTM International Committee E60 on Sustainability is applying the principles of sustainability through standards development. This effort is pioneering a comprehensive life cycle approach that includes the more familiar environmental consideration of ?going green? by recycling and reducing waste, plus the addition of economic and social principles. One of the goals is the development of standards to meet the sustainability needs of the present without compromising the ability of future generations to meet their own needs.

The standards development activity was organized by ASTM in October 2008 and involves both existing and new standards activities that will encourage sustainability practices. Initial activties have focused on topics such as Green Meetings and Buildings Products. One proposed new ASTM standard covering the Classification for Environmentally Preferable Products provides a framework for a consistent approach to the format and content of environmental information and in turn will enable the user to evaluate the comparative environmental impact of building products. The proposed classification provides the first comprehensive standard in this area, and that the April meeting attendees will discuss results of a pilot program to refine the document and verify its viability in the marketplace. Additional ASTM standards development activties includes proposed standard practices for water-conserving systems, test methods related to assessing aspects of green roofs, and a draft Standard Guide for Marketing and Product Claims Related to Sustainable Building.

In summary, ASTM International Committee E60 provides a global forum where diverse stakeholders come together to create the consensus standards that will promote and integrate sustainable development across multiple industry sectors.

**Science-Based Guidance for Environmental and Economic Sustainability**

Lippiatt, B., NIST

Originally presented on: 8/12/2009 15:00:00 - 15:25:00

Industry stakeholders need compelling metrics, tools, data, and case studies supporting major investments in sustainable technologies. Proponents of green building, for example, widely claim that these buildings are cost-effective, but often these claims are based on incomplete, anecdotal evidence that is difficult to reproduce and defend. The claims suffer from two main weaknesses: (1) technologies upon which claims are based are not necessarily ?green? in a science-based, life-cycle assessment (LCA) sense and (2) their measures of cost-effectiveness often are not based on standard methods for measuring economic worth. Yet industry demands compelling metrics to justify sustainable technology investments. The problem is hard to solve because, until now, neither methods nor robust data supporting defensible business cases were available. The National Institute of Standards and Technology (NIST) is beginning to address these needs by developing rigorous metrics and tools for scientifically assessing the life-cycle economic and environmental performance of complex industrial systems such as buildings, automobiles, and farm equipment. Economic performance is measured using standard life-cycle costing methods. Environmental performance is measured using LCA methods that assess the ?carbon footprint? of industrial systems as well as 11 other sustainability metrics including fossil fuel depletion, smog, water use, habitat alteration, indoor air quality, and human health. Carbon-efficiency ratios and other eco-efficiency metrics are established to yield science-based measures of the relative worth, or ?business case,? for sustainable technologies. The approach is illustrated through case studies of energy efficiency investments in commercial buildings.

**ANSI Green Chemical Standard**

Peoples, R., ACS Green Chemistry Institute
Young, J., ACS Green Chemistry Institute
Originally presented on: 8/12/2009 15:45:00 - 16:10:00

The ACS Green Chemistry Institute® has initiated a program for the development of an ANSI standard for green chemicals and green process technology. This talk will outline that initiative and give a status report on the current state of development of this new consensus standard. A large stakeholder group is working to create a standard that is similar in design to the LEED certification for green building in that it will contain categories for performance along with specific credits. These credits will generally cover the following impact area: public health and environment, energy, water, manufacturing operations, end of life considerations and social equity. The effort is designed to bring a new level of clarity and performance to the chemical enterprise and provide a tool for users to guide decisions. Whether you are trying to find the right chemical for an application or decide on the best manufacturing route, the new standard will provide guidance.

# Nanoparticles: New Opportunities That Are Facing EHS Concerns and Regulatory Uncertainties

## Authors:

Carberry, J. B., Dupont

Recent advances in the ability to make engineered nanoparticles and incorporate them into a range of fibers, films, engineering plastics and electronic materials has given rise to many new product opportunities. Along with these product opportunities, potential EHS concerns have arisen in an atmosphere of regulatory uncertainty on dealing with these materials. Working with Environmental Defense Fund, the DuPont Company developed a Framework for guiding Product Stewardship of nanoparticles. Practical application of that Framework will be reviewed along with findings on needs and availability for Stewardship information.

**Codes and Standards: Enablers for Sustainable Growth**

Taubitz, M. A., Lean Journey LLC

Originally presented on: 8/12/2009 16:35:00 - 17:00:00

This presentation will focus on how codes and standards are enablers for injecting science into the deliberations on sustainability. Codes and standards are foundations for the design of new products and processes. Developed using a formal consensus process, codes and standards offer the opportunity for all stakeholders to vet and balance sustainability issues with other requirements. This open and balanced development period is a time when professionals and technical experts can bring data and credible evidence to the table. This would inject science and allow balancing issues affecting people and planet against performance, cost, safety and other design criteria. Such a proactive approach would be invaluable in guiding companies and organizations dedicated to sustainable growth.

The term ?standard? has a very different meaning to different stakeholder groups. This issue is explored to enhance understanding. In the business world, ?standards? are part of lean production, which has been proven to reduce environmental impacts. EPA and industry have found ?lean? to be a significant enabler of ?green? and sustainability.

# A Transdisciplinary, Transinstitutional, and Transnational Approach to Urban, Civil Infrastructure Systems, and Facilities Sustainability


## Authors:

Vanegas, J., Texas A&M


21st century challenges require new bold, innovative, systems based, and contextually sensitive approaches to the development of solutions for a Sustainable Built Environment (SBE) in general, and more specifically, for Sustainable Urbanism, Civil Infrastructure Systems, and Facilities (SUCISF) solutions, given the interrelationships, interdependencies, and complexity of the external factors affecting their delivery. Five key imperatives drive the response to these challenges. First, governments, together with other stakeholders in the public and private sectors, must work collaboratively at local, national, regional, and international levels to pursue integrated SBE/SUCISF solutions. Second, public and private sector initiatives to plan, finance, develop, and deliver these types of SBE/SUCISF solutions need to address formally, explicitly, and proactively, in an integrated

way, the three dimensions of sustainability (environmental, social, and economic), as well as the

elimination, reduction, and mitigation of risk and vulnerability of the built environment to natural

hazards. Third, these initiatives also cannot afford to continue following the same strategies,

mechanisms, and processes that have been used to date. Fourth, government officials, policy makers,

regulatory agencies, finance institutions, public and private owners of civil infrastructure systems and

facilities, community leaders, planners, architects, landscape architects, engineers, suppliers,

constructors, or end-users cannot overcome these challenges alone. Finally, fifth, all stakeholders

must link, coordinate, and integrate their efforts as a single cohesive critical mass, pooling,

leveraging, and sharing their resources, within public/private partnerships at any level, from local to

international, in the pursuit of SBE/SUCISF initiatives of common interest and benefit to all.

This presentation will discuss a current initiative to address these challenges and respond to these

imperatives: a transdisciplinary, transinstitutional, and transnational approach to urban, civil

infrastructure systems, and facilities sustainability. This approach to SBE/SUCISF (1) aims to create

sustainable urban places by synthesizing and filling the gaps between sustainable development and

urban planning, design, and construction at a broader level; and more specifically (2) explores

sustainability by focusing on the design, form, function, and beauty of the built environment ? i.e.,

land developments, built landscapes, civil infrastructure systems, and facilities ? that collectively

make up human settlements at various spatial and temporal scales. The presentation will discuss

fundamental principles for sustainable urbanism, and basic strategies for the planning, design,

procurement, construction, and commissioning of civil infrastructure systems and facilities, which can

be applied to achieve built environment sustainability as one of the means for satisfying the critical

needs of society today, both in the U.S. and internationally

**Trends in Sustainable Design and Architecture**

Fillmore, R., Dow Corning

Originally presented on: 8/12/2009 14:10:00 - 14:35:00

In April 2009, the U.S. Green Building Council announced that the overall green building market is likely to more than double in size by 2013. Dow Corning is designing innovative technologies to meet the needs of industry experts by delivering next-generation products, process advances and solutions. In this presentation, Ron Fillmore will highlight the latest trends in the industry and will offer insights on ways that materials suppliers, architects and the construction industry can work together to develop sustainable solutions.

# Measuring Sustainability

## Authors:

Hansen, V., U.S. EPA

The definition of ?sustain' is to keep something in existence. It is this objective and definition that provide some insight into the metrics that are required to measure sustainability and adequately respond while something is still viable. Indicators do not always adequately measure sustainability because they provide answers to whether sustainability was achieved. If the answer is ?no', they have done little to ?keep something in existence' and can not undo prior losses. Keeping something in existence requires metrics that guide human decisions and activities that could threaten losses so that such losses are avoided. Because intact natural, social, and economic systems are essential to sustain human life, keeping these systems intact requires that human actions/behaviors and development move toward harmony with these systems. Therefore, sustainability metrics must directly address human objectives and environmental, social, and economic constraints that influence meeting these objectives.

**Measuring Sustainability of Alternatives to Use of Potable Water to Flush Toilets Using Life Cycle Assessment**

Apul, D., University of Toledo
Anand, C. K., University of Toledo
Originally presented on: 8/12/2009 15:00:00 - 15:25:00

The drinking water and wastewater infrastructures in the US are aging and are unable to square the mounting pressures due to degradation of pipes. There is a serious shortfall of funds between current investments and the investments that will be needed, in the next few decades, to meet mandates of Clean Water and Safe Drinking Water Acts. In addition, the current water infrastructure is very energy intensive which makes the entire process even more expensive. One solution for addressing these issues is by implementing integrated water management systems in buildings that use less potable water and send less water to sanitary and storm sewers, to benefit from the reduction of energy use to treat the amount of water being sent to the waste water treatment plants. Toilet flushing is one of the large applications of potable water. Implementation of the most sustainable alternative to use of potable water for flushing toilets ,to a large set of buildings in an area, can significantly help reduce the energy use, cost and also carbon emissions over a period of time in that area. There has been much renewed interest in this topic partially because of the popularity of the US Green Building Council's LEED rating system that includes water efficiency and sustainable sites credits to address storm, potable, and sanitary sewer water quantities. Currently, designers pick supposedly sustainable solutions for a specific building in absence of scientific data on the actual impacts and benefits of the proposed solution. The focus of this research is comparison of four alternatives to use of potable water in flushing toilets in North Engineering building at the University of Toledo. The alternatives compared are i) use of roof water to flush toilets, ii) Use of grey water to flush toilets iii) using composting toilets, iv) use of low flush toilets, based on the following criteria, a) amount of potable water being used b) energy use for operation of these systems c) Cost and energy spent on treating this amount of water, in each case, using life cycle assessment. Sustainability indicator analysis and also, the possibility of obtaining LEED credits by using each of the methods will be determined.

**Comparative Life Cycle Assessment of Insulating Concrete Forms with Traditional Residential Wall Sections**

Rajagopalan, N., University of Pittsburgh
Bilec, M., University of Pittsburgh
Landis, A. E., Arizona State University
Originally presented on: 8/12/2009 15:45:00 - 16:10:00

Buildings account for 30-40% of the energy use. In United States, buildings use 70% of the nation's electricity and emit 40% of the greenhouse gas emissions (GHGs). Green buildings and green building products are increasingly becoming the norm in the construction industry. But there is no consensus on labeling of a green product. There is a need for a holistic system which looks at all aspects of a material or product and then labels it green. This research incorporates the knowledge of building construction and sustainability tools, such as life cycle assessment (LCA), to compare traditional and new building materials and to analyze whether the new building products are truly green and can be used as replacements for the traditional building materials in certain structural components.

Insulating Concrete Forms (ICF) is a building material which is made of polystyrene into which concrete is poured. Unlike conventional construction, the forms are not removed after concrete has cured and act as insulation. Ties made of polypropylene act as reinforcement for the material and additional steel reinforcement can be added where necessary. In this research, this building material is compared with traditional materials like wood using LCA.

Previous studies have compared ICF with traditional materials. A comparative partial life cycle inventory was conducted on 2,450 square feet houses made of ICF and wood. The inventory was conducted at five different locations in the United States- Phoenix, Miami, Seattle, Washington,DC and Chicago. The results of the study show that the initial embodied energy of ICF is higher than wood but the cumulative energy after five years is higher for wood than ICF. The use phase accounts for 99% of the total energy consumption whereas the cement and concrete production account for less than 1% of the total energy consumed. Another study which compared 2,400 square feet houses made of ICF, wood and steel frame stated that the ICF had higher embodied energy than the other two materials. The National Association of Home Builders (NAHB) constructed demonstration homes made of ICF to show people about the alternative materials available in the market and the method to use them. The desirable properties such as serviceability, energy efficiency and durability compensate for the high initial price. An extension of the partial life cycle inventory study was an energy use study conducted in the same locations as mentioned. Energy modeling tools were used analyzing the energy consumption of the building in different locations. The results of the study showed that ICF walls have an inherent capacity of higher insulation than wood framed walls and have 5% to 9% greater energy savings than the wood framed walls.

While there are many benefits of ICF, a detailed study of the material in comparison with traditional materials is required to identify the advantages and disadvantages of the material in its life cycle. Before ICF is labeled as a ?green building product?, the impact of the product in its life cycle must be analyzed.

The goal of this study is to perform a comparative LCA of residential structures using new and traditional building materials. The entire life cycle of the residential building is divided into four phases- manufacturing of raw materials, construction of the structure, use of the structure and end of life of structure. This research has currently modeled the manufacturing phase and the use phase of the structure. Previous residential LCAs have taken the entire house as the functional unit for all phases which shows that the use phase is the highest contributor to environmental impact. If a complete understanding of the various phases in the life cycle is required, using a single functional unit for all phases does not provide a clear picture. Extensive data is required for every project if the whole house is taken as a functional unit which prevents an unclear picture of the true environmental hot spots. This research aims to create a residential model which will identify areas in all the phases of the life cycle which need to be addressed instead of concentrating on just the use phase. For the manufacturing phase of the structure, a functional unit of one square feet of the structure is assumed and the quantity of raw materials required for producing the functional unit is calculated. Data is obtained from a variety of databases like Ecoinvent, IDEMAT 2001 and ETH-ESU and combined and the input along with the material quantity in the modeling tool, SimaPro to obtain an inventory. Data from industry was also collected and used in the model.

For the use phase, the use of the materials in a 2, 450 square feet house is compared. Various scenarios which involves houses purely made of the aforementioned building materials and a combination of the building materials in different structural components are modeled. The basic features of foundations, windows, doors remain the same for all houses. The modeling tool used for this phase is a Department of Energy (DOE) freeware called eQuest.

Preliminary results show that in the manufacturing phase, ICF does not perform as well as wood in environmental impacts. But in the use phase, a house made of a combination of wood and ICF has the minimum energy use. Any house modeled with a combination of ICF performs better than a house purely made of traditional building materials. Subsequently, the less energy use by a combination of wood and ICF house translates to lower electricity bills annually and lower carbon dioxide emissions per year.

A more robust LCA which includes both process based and input-output data needs to be performed in the remaining phases of the life cycle before a decision can be made on the environmental impacts of the material. As we are moving towards a world which faces energy crisis and expensive energy solutions, materials which can reduce the use phase costs become increasingly necessary. LCA results show the hotspots associated in the manufacturing of the raw materials which need to be addressed with alternative strategies. The desirable qualities of ICF should be tapped after a thorough study of the material has been conducted to address the problem of the life cycle use phase impacts of buildings.

**Integrating Design for Assembly and Design for Disassembly in Life Cycle Management for Developing a Sustainable Product Development Design Methodology**

Memary, R., Universiti Teknologi Malaysia
Mat Saman, M. Z., Universiti Teknologi Malaysia
Bin Sharif, S., Universiti Teknologi Malaysia
Originally presented on: 8/12/2009 16:10:00 - 16:35:00

The extremely high and ever-increasing annual disposal rates of solid waste have caused a big problem for environmental protection in the world. Human society consumes resources, but it is not able to reproduce them. Human activities are still based on ??open cycles," starting from a condition of natural environmental balance and reaching an environmental imbalance. The challenging scope of scientific and technological research towards sustainability appears clear if it is based on this analysis: to find development systems based on ??closed cycles" of resources. In order to achieve closed cycles during product design and development process, those stages that are related to sustainability need a new methodology. Between product development process stages, assembly and disassembly play key roles in supporting sustainability. Design for assembly is an approach to design the product in the way that has least number of parts and simplest assembly process with minimum cost required. And design for disassembly is an approach to find the best disassembly sequence with least time. But newly there are other approaches implemented in design for disassembly including approaches that influence sustainability. As a case in point enabling disassembly to support maintenance and part recycling is current research issue. There are different methodologies in design for assembly including design principles and rules, quantitative evaluation procedures, knowledge based approach, and computer aided design for assembly. On the contrary there are less developed procedures for design for disassembly. There are some quantitative and some knowledge based approaches developed for design for disassembly. In this paper, different methodologies in design for assembly and design for disassembly based on four case studies including a hair clipper, a speaker, an electric pot and a printer are studied and stages which are related to sustainability are highlighted. Next, regarding the relationship between those stages and cost analysis in product lifecycle management a new methodology that combines design for assembly and design for disassembly in product design stage is proposed. This methodology considers product recyclability in the early stage of design. Although this methodology may increase the assembly cost of some components, but totally leads to a better justification for product End of Life Value. This approach designs sustainable products which end to less disassembly cost and justifies the part recycling.

**Design for Disassembly in the Built Environment**

Guy, G. B., Carnegie Mellon University

Originally presented on: 8/12/2009 16:35:00 - 17:00:00

Design for disassembly (DfD) is a growing topic within manufacturing industries as greater attention is devoted to the management of the end-of-life of products. This need is driven by the increasing disposal problems of large amounts of consumer goods, and the resultant pollutant impacts and the loss of materials resources and energy that is embodied in these products. In Europe, in particular, constraints on the production of waste and pollution have resulted in an emphasis on ?extended producer responsibility? legislation such as the Directive 2000/53/EC of the European Parliament on end-of-life vehicles and Germany's End-of-Life Vehicle Act of 2002.Extended producer responsibility (EPR) is defined as ??making the manufacturer of the product responsible for the entire life-cycle of the product and especially for the take-back, recycling and final disposal of the product? (Toffel, 2002). Simply put this requires that those who create a product are responsible for designing its entire life-cycle, including its ultimate disposition, with reuse and recycling, to achieve economic profitability at minimum risk. As buildings are manufactured artifacts typically comprised of a combination of pre-assembled components and on-site assembly of materials and components, design for disassembly can be applied to the built environment similarly to other assembled artifacts. Although not typically considered a ?product?, buildings are nonetheless composed of materials, components, and connections, and are created through the collaboration of designers, engineers, trade-persons, and the manufacturers of the materials and components that are assembled to make the building. Buildings of course are much longer-lived than other artifacts, larger, more complex and subject to a myriad of forces, such as gravity, climate, and use. Because of their importance in society and tremendous impact on resource utilization globally, any attempt to consider sustainability in the use of building-related resources must consider the management of all resource flows in the full life-cycle of buildings from extraction, to manufacturing, to design, to construction, to operation, to renovation, to eventual end-of-life. This management is a fundamental goal of design for disassembly (DfD) in the built environment. There are many practical reasons for incorporating DfD in the built environment. The US Geological Survey has estimated that 60% of all materials flow (excluding food and fuel) in the US economy is consumed by the construction industry (Wagner, 2002). The US EPA has estimated that 92% of all construction-

related waste produced annually in the US is the result of renovations and demolitions, with only 8% produced from new construction, and that this waste is upwards of 30% of all waste produced in the US (Franklin Associates, 1998). Nelson has estimated that the total built space in this country will need to grow from 296 billion square feet in 2000, to 427 billion square feet in 2030. Of this growth, 82 billion square feet of building will be from replacement of existing building space and 131 billion will be from new construction totaling 213 billion square feet of new built space. This means that 27% of existing buildings in the year2000 will be replaced from 2000 to 2030 and that over 50% of buildings in the year 2030 will have been built since 2000 (Nelson, 2004). This huge mass of buildings that are to be replaced and newly constructed can either be large sources of waste in the next 25 years and afterwards, respectively, or the materials from existing buildings can be recovered for reuse and recycling, and these new buildings canbe designed to recover their respective materials in the next generation of replacements. DfD is intended to address this trend by explicitly creating buildings to reduce new materials consumption and waste in their construction, renovation and demolition, to increase building lives in situ, and to create buildings that are stocks of future building materials.This enabling of materials conservation and buildings that facilitate the recovery of their components for the next iteration is intended to provide both economic and environmental benefits for builders,owners and occupants, and communities where these buildings reside.

# AUTHOR INDEX

# AUTHOR INDEX